Climate Change Education

Climate Change Education

Reimagining the Future with Alternative Forms of Storytelling

Edited by Rebecca L. Young

Afterword by Vandana Singh

LEXINGTON BOOKS
Lanham • Boulder • New York • London

Published by Lexington Books
An imprint of The Rowman & Littlefield Publishing Group, Inc.
4501 Forbes Boulevard, Suite 200, Lanham, Maryland 20706
www.rowman.com

86-90 Paul Street, London EC2A 4NE

British Library Cataloguing in Publication Information Available

Library of Congress Cataloging-in-Publication Data Available

ISBN: 978-1-66691-579-2 (cloth)
ISBN: 978-1-66691-580-8 (electronic)
ISBN: 978-1-66691-581-5 (pbk.)

Contents

Acknowledgments   vii

Introduction   1

Chapter One: Reading the Youth Climate Movement: Social Media,
Literary Creation, and Allyship   9
Alexandra Lakind

Chapter Two: "But, What Difference Can I Make?": Using
Documentaries to Explore Environmental Advocacy in the Face
of Climate Change   29
Carley Petersen Durden and Jared Durden

Chapter Three: Educating Space-Age Environmentalists at the
Elementary Level   49
Beverly B. Bachelder and Robert S. Bachelder

Chapter Four: Teaching Environmental Respect to Young Learners:
Video Games as Environmental Texts   75
Erden El

Chapter Five: A City for the Future: Designing Socially Just,
Sustainable Urban Environments with Elementary Students   95
Alexandra Laing

Chapter Six: Fostering Environmentalism and Activism in Students:
Plastic Pollution as a Starting Point   113
Karen Ball and Elke de Vries

Chapter Seven: Ecohorror, Terrorism, and Inadequate
Representation of Global Warming in M. Night Shyamalan's
The Happening   137
Tatiana Konrad

Chapter Eight: The Global Impact of Fast Fashion: Understanding
  Sustainability and Social Justice Issues   151
  Helen Liu and Alyssa Racco

Chapter Nine: Making the Material Turn: A Pedagogical Approach
  on Postcolonial, Social, and Ecological Issues in Amitav Ghosh
  and Arundhati Roy's Essays and Fiction   177
  Suhasini Vincent

Chapter Ten: Creating Authentic Learning Experiences:
  Interdisciplinary Climate Change Instruction and Assessment   201
  Mary-Alice Corliss and Rebecca L. Young

Afterword   231
  Vandana Singh

Index   235

About the Contributors   241

Acknowledgments

Thank you to all who have supported this project with encouragement, advice, and patience. I am grateful to the incredibly dedicated educators who have shared their ideas and guidance in this collection. Climate literacy begins in the classroom, and every teacher has the power to inspire student awareness, readiness, and hope. We invite readers to join this awesome effort!

Introduction

"Two astronomers go on a media tour to warn humankind of a planet-killing comet hurtling toward Earth. The response from a distracted world: Meh."[1] The 2021 Netflix film *Don't Look Up* would be hilarious if it weren't such a frightening parody of our time. The astronomers' experiences attempting to inform the world of imminent crisis and to convince the American president that definitive action is needed are perhaps exaggerated but still poignant rebukes of how policy makers and the media have failed to take seriously the warnings of climate scientists. In the film, even those who do fight to spread awareness help illuminate director Adam McKay's message about our collective failure to act with resolve on climate change. As *New York Times* film critic Manohla Dargis puts it,

> McKay's work with DiCaprio is particularly memorable, partly because Dr. Mindy's trajectory—from honest, concerned scientist to glib, showboating celebrity—strengthens the movie's heartbreaking, unspeakable truth: Human narcissism and all that it has wrought, including the destruction of nature, will finally be our downfall. In the end, McKay isn't doing much more in this movie than yelling at us, but then, we do deserve it.[2]

Indeed, Jennifer Lawrence's character (as the PhD student who initially discovers the comet) does yell at the audience in several scenes, attempting to make their findings explicitly clear: the comet will destroy Earth if nothing is done to stop it (a powerful filmic contrast to the sometimes more ambiguous ways climate science has been presented to the general public). That the majority of characters either ignore or mock her even as Dr. Mindy enjoys shallow fame from their appearances speaks to another critique of reactions to climate science that are rooted in fear, misunderstanding, and denial. DiCaprio notes that the intent behind Dr. Mindy's character was "to emulate the frustrations of the scientific community, especially scientists that have felt pushed aside and marginalized and have tried to convey the science and the

1

truth about the time frame that we have to solve this crisis . . . "[3] But what NASA climate scientist James Hansen tried to make Congress understand in 1988 about the climate crisis—"It is already happening now"—has still not quite grabbed hold. Decades later, Hansen, now a vocal climate activist, regrets that he was not "able to make this story clear enough for the public."[4]

Stories about climate change help to clarify what may yet seem abstract about the science. In the film, clarity emerges in the form of a giant comet becoming visible in the atmosphere, prompting division in the public and two ridiculously distinct campaigns: "Don't Look Up" versus "Just Look Up." It turns out there isn't time to "sit back and assess" the "boring" details of planetary catastrophe that Dr. Mindy desperately tries to convey to the U.S. president and her team—the comet's impact is imminent. It's happening and the storyline offers no realistic exit strategy for the characters. For the audience, let us hope that clarity comes sooner.

In *Affective Ecologies: Empathy, Emotion, and Environmental Narrative*, Alexa Weik von Mossner explains the tangible effect on audiences of another climate crisis film: "a study conducted by Yale psychologist Anthony Leiserowitz reveals that *The Day After Tomorrow* generated 'more than 10 times the news coverage of the 2001 [Intergovernmental Panel on Climate Change] IPCC report.'"[5] Leiserowitz's summary of the American study revealed that

> the film led moviegoers to have higher levels of concern and worry about global warming, to estimate various impacts on the United States as more likely, and to shift their conceptual understanding of the climate system toward a threshold model. Further, the movie encouraged watchers to engage in personal, political, and social action to address climate change and to elevate global warming as a national priority. (2004, 34)[6]

Despite the obvious fictionality of the film, viewers reacted with real consequence because its characters and storyline impacted them on an emotional level, a perspective-taking effect that scientific graphs and data simply may not prompt. Put another way, until we can see the comet speeding toward us, there seems to be room for doubt. While McKay's *Don't Look Up* has faced some critical film reviews, its rather unhappy ending has generated some much-needed news coverage and interviews about climate change.

An interesting contrast in how both films address leadership's failure to prevent disaster is in the fates portrayed for the American presidents. Weik von Mossner explains of *The Day After Tomorrow*, "the greatest personal transformation and learning experience" is observed in the character playing the new president of the United States:

In his first TV address to the nation after the disaster, he displays a new sense of humility and environmental ethics. "For years," he states, "we operated under the belief that we could continue consuming our planet's natural resources without consequences. We were wrong. I was wrong. The fact that my first address to you comes from a consulate on foreign soil is [testimony] to our changed reality." This is the narrative's moment of catharsis and recognition. . . . [7]

In *Don't Look Up* the audience is not granted such a satisfying realization, though they may find the resolution offered more gratifying in the moment. Meryl Streep's role as president is a scathingly humorous portrayal of hubris and power. Her character, who abandons Earth on a futurist's backup escape plan ship to the future, is dealt with more like the deceptive politicians and wealthy supporters of a similarly unjust climate scheme in *Kingsman: The Secret Service*. She loses her head.

More interesting is how the films craft the fate of the other characters. In *The Day After Tomorrow*, "the United States and the rest of the world are permanently transformed and order is not reestablished. Only utter devastation, Emmerich seems to suggest, can finally give humanity the ability to learn from its mistakes and change its unsustainable practices."[8] At least they have a chance. McKay takes a different approach—the comet destroys Earth, all the story's likable characters, and (one can presume) the survivors who land in the future will meet an ending not too unlike their leader's. The *New York Times* review suggests that maybe we deserve this, but perhaps the humor throughout this film is its triumph. There are no heroes to save the day at the final tender moments of the astronauts' story, but one can at least hope that viewers will just look up in time.

As Jonathan Gottschall details in *The Storytelling Animal: How Stories Make Us Human*, there is a rather persuasive explanation behind why these climate crises films and other such "fictions" have the potential to sway public opinion:

Fiction is a powerful and ancient virtual reality technology that simulates the big dilemmas of human life. When we pick up a book or turn on the TV—whoosh!—we are teleported into a parallel universe. We identify so closely with the struggles of the protagonists that we don't just sympathize with them; we strongly empathize with them. We *feel* their happiness and desire and fear; our brains rev up as though what is happening to them is actually happening to us.[9]

He summarizes of the science behind this level of impact that "research results have been consistent and robust: fiction *does* mold our minds. Story—whether delivered through films, books, or video games—teaches us facts about the world; influences our moral logic; and marks us with fears, hopes, and anxieties that alter our behavior, perhaps even our personalities."[10]

This level of empathic engagement is of particular interest to educators who wish to put good stories in service of learning. Now more than ever schools everywhere should be preparing students not only for roles in their local community but for their participation as global citizens. We do an incredible disservice to students if we allow them to graduate without an understanding of climate change, including the intricate contexts that have brought us to this point of urgency and the severe social and environmental injustices its impacts exacerbate. If we want to graduate twenty-first-century learners, we must educate based on the real challenges we face, recognizing that the solutions our students ultimately seek will define the path forward for us all. Climate education is an opportunity to help young people create a better future.

A well-designed K–12 climate education provides not only the content students need to understand climate change and the patterns of human behavior that contribute to it but the humanitarian lens through which they might contextualize its many consequences. It also offers the interdisciplinary tools students must be armed with to envision solutions that help mitigate worst-case scenarios.[11] As youth activists around the world are declaring, this educational imperative will define their future. Facing climate change means confronting the gross inequities that plague us. It means managing myriad crises that already haunt our collective conscience. It means prioritizing the planet while we still have a short window of time to stave off global collapse.

Significant opportunities exist for immediate actions we might take from green energy and engineering solutions to technological innovations we cannot yet imagine. Whatever short-term expenses these actions incur, they are miniscule in contrast with the steep and lasting costs of inaction. While teachers certainly do not deserve to have one more expectation laid upon them, the unique nature of our time demands it: education will define the path ahead.

This collection seeks to support educators who are ready to answer students' call for comprehensive climate education—an initiative already well underway in classrooms around the world. We hope the following chapters will provide guidance for building on this momentum and offering climate education across grade levels and subject areas.

Highlighting ways leaders of the Youth Climate Movement have harnessed the power of digital storytelling to organize protests that demand climate action, "Reading the Youth Climate Movement" offers educators ideas for learning with students in this unique space. Through writing exercises based on social media content, Alexandra Lakind demonstrates how student engagement on platforms like Instagram can both inform their understanding of climate change and empower their desire to confront it.

In "'But, What Difference Can I Make?': Using Documentaries to Explore Environmental Advocacy in the Face of Climate Change," Carley Petersen Durden and Jared Durden relate their experience using the films *Bugs* and *Flat Earthers: Behind the Curve* in first-year college classes to foster conversations about the influence of biases and misinformation. Analyses based on the film studies serve as entry points for discussing skepticism about climate change.

"Educating Space-Age Environmentalists at the Elementary Level" pairs storytelling and space data in a creative approach designed to help students understand the importance of responsible space exploration. Advocating for the role of space literacy in climate education, Beverly B. Bachelder and Robert S. Bachelder offer pedagogical strategies that encourage student agency for cleaning up the environment both on Earth and in space.

"Teaching Environmental Respect to Young Learners: Video Games as Environmental Texts" explores how to engage students in ecocritical studies through gaming. Erden El presents background on the approach, examines DeLillo's *White Noise* as a model for ecocritical analyses, and outlines a similar approach based on two video games that engage young people in climate activism.

"A City for the Future: Designing Socially Just, Sustainable Urban Environments with Elementary Students" describes a year-long interdisciplinary inquiry project that allows students to investigate multiple aspects of city design. With *The City of Ember* as an anchor text, Alexandra Laing offers opportunities for students to explore environmental and social justice concerns related to urban areas and innovate sustainability solutions in a city of their creation.

"Fostering Environmentalism and Activism in Students: Plastic Pollution as a Starting Point" explores the role of poetry, short stories, and documentaries in protesting issues related to environmental pollution, particularly plastics. Karen Ball and Elke de Vries offer an instructional approach appropriate for middle or high school students that encourages ecocritical approaches to literature and film studies, fosters the practice of persuasive language skills and techniques, and culminates in action-based community research and problem-solving activities.

"Ecohorror, Terrorism, and Inadequate Representation of Global Warming in M. Night Shyamalan's *The Happening*" offers an ecocritical analysis of the film and its comparison of global warming to the 9/11 terrorist attacks. Drawing parallels between cultural imaginings of the attack and climate change, Tatiana Konrad explores new directions in ecohorror while offering strategies for engaging upper-level students in Shyamalan's portrayal of plants as terrorists and the film's themes of hostility, mutation, and vengeance.

Examining the relationship between fashion advertising and textile manufacturing, "The Global Impact of Fast Fashion: Understanding Sustainability and Social Justice Issues" details the far-reaching consequences of this industry. By telling the story of fast fashion, Helen Liu and Alyssa Racco help young people understand the true cost of consumer culture and its impact on children, women, and the environment. Armed with knowledge, students are encouraged to fight for systemic change by demanding transparency and responsibility of the brands they support.

In "Making the Material Turn: A Pedagogical Approach on Postcolonial, Social and Ecological Issues in Amitav Ghosh and Arundhati Roy's Essays and Fiction," Suhasini Vincent explains how stories about climate change can inform students' understanding of its interrelated causes and consequences, particularly in developing parts of the world. Illustrating the importance of analyzing global problems through the lens of postcolonial ecocriticism, this chapter provides guidance on how to engage students in this effort and support their awareness of systemic injustice.

The final chapter, "Creating Authentic Learning Experiences: Interdisciplinary Climate Change Instruction and Assessment" offers a rationale for using narratives as a lens for teaching climate science and fostering empathy. Mary-Alice Corliss and Rebecca Young argue that we meet students' demand for comprehensive climate education by aligning pedagogy to the twenty-first-century skills students will need to overcome its challenges.

We hope *Climate Change Education* will support teachers in classrooms around the world who are inspiring students to imagine a better future.

NOTES

1. Adam McKay, *Don't Look Up* (2021; New York City: *Netflix*) Streaming, https://www.netflix.com/title/81252357.

2.2. Manohla Dargis, "'Don't Look Up' Review: Tick, Tick, Kablooey," *New York Times*, December 23, 2021, https://www.nytimes.com/2021/12/23/movies/dont-look-up-review.html.

3. David Canfield, "How *Don't Look Up* Assembled Its Insanely Starry Cast," *Vanity Fair*, January 27, 2022, https://www.vanityfair.com/hollywood/2022/01/awards-insider-dont-look-up-ensemble-feature.

4. Seth Borenstein, "James Hansen Wishes He Wasn't So Right about Global Warming," AP News, June 18, 2018, https://apnews.com/article/science-30-years-of-warming-us-news-climate-james-hansen-664cf2e917604adf90472daa35989ffb.

5. Weik Von Mossner, *Affective Ecologies: Empathy, Emotion, and Environmental Narrative*, (Columbus: The Ohio State University Press, 2017),158.

6. Quoted in Weik Von Mossner, 159.

7. Weik Von Mossner, 157.

8. Weik Von Mossner, 158.
9. Jonathan Gottschall, *The Storytelling Animal: How Stories Make Us Human.* (First Mariner Books, 2013), 67.
10. Jonathan Gottschall, 148.
11. Rebecca Young, *Confronting Climate Crises through Education: Reading Our Way Forward.* (Lanham: Lexington Books, 2018).

BIBLIOGRAPHY

Borenstein, Seth. "James Hansen Wishes He Wasn't So Right about Global Warming." AP News, June 18, 2018. https://apnews.com/article/science-30-years-of-warming-us-news-climate-james-hansen-664cf2e917604adf90472daa35989ffb.

Canfield, David. "How *Don't Look Up* Assembled Its Insanely Starry Cast." *Vanity Fair*, January 27, 2022. https://www.vanityfair.com/hollywood/2022/01/awards-insider-dont-look-up-ensemble-feature.

Dargis, Manohla. "'Don't Look Up' Review: Tick, Tick, Kablooey." *New York Times*, December 23, 2021. https://www.nytimes.com/2021/12/23/movies/dont-look-up-review.html.

Gottschall, Jonathan. *The Storytelling Animal: How Stories Make Us Human.* First Mariner Books, 2013.

McKay, Adam. *Don't Look Up.* 2021. New York City. *Netflix*. Streaming.

Weik Von Mossner, Alexa. *Affective Ecologies: Empathy, Emotion, and Environmental Narrative.* Columbus: The Ohio State University Press, 2017.

Young, Rebecca. *Confronting Climate Crises through Education: Reading Our Way Forward.* Lanham: Lexington Books, 2018.

Chapter One

Reading the Youth Climate Movement

Social Media, Literary Creation, and Allyship

Alexandra Lakind

In recent years youth activism has become a formidable force in the global social and political landscape. Specifically, the School Strikes for Climate (SSC) are reshaping environmental priorities and providing new frameworks, logics, and directions for action on climate change. In so doing, I will suggest, these movements expand and challenge existing approaches and structures to climate education, shifting how and where learning occurs. Using social media as a primary tool to organize and communicate, youth activists proffer a view of climate change as a crisis of intersecting unsustainable systems that cannot be addressed with a narrow focus on carbon in the atmosphere. Even without geographical, ideological, or cultural consensus, these strikes (and the organizing around them) offer the broader climate movements a chance to learn about the power of youth and the interconnected issues of climate change and justice.

As youth activists confront the inadequate ways prior movements have attended to issues of justice, they've scrutinized mainstream Western environmental messages. For example, numerous authors have critiqued slogans such as "we are all in this together." As environmental studies theorist Rob Nixon reminds,[1] we may be in this together, but not all in the same way. For instance, as the impacts of climate change intensify over time, young people will experience greater consequences than older generations. Further, contemporary champions of climate justice reject the once-popular vision of the universal "we" wherein everyone is implicated, and where the "every-one" is too often depicted as Western, white, and middle class, erasing the complex and intertwined histories of colonialism and capitalism that manifest in present-day issues of environmental injustice.[2] In a sobering analysis, the youth climate movement calls out how adults, who promised to protect them, have burdened them with the impossible task of complete global economic and social restructuring. More than just criticizing adults, the youth climate movement has sought to educate and mobilize in ways that overcome the failures of previous environmental movements.

There is a new canon of literature showing the vast interconnectedness between issues of climate change and social justice emerging, thanks to countless new texts published and shared through social media. This canon is being determined socially and collectively. One post, for example, teaches the reader about Indigenous justice simply by listing accounts that one might follow alongside the text: "make time to learn about the stolen land you're on." This chapter asks educators to consider integrating these texts into their lesson plans. The texts are short, but the form varies widely from 280-character tweets to hashtags written atop images. Figure 1.1 is a post created by Métis artist Kim Stewart of her countermapping practice where she uses beadwork to challenge settler assumptions represented through objects such as economic maps that delineate space based on notions of property. This image was reshared by youth activists with the accompanying text: #landback, a hashtag popularized in 2019 premised on a movement to return land to Indigenous peoples, protect Indigenous knowledge of environmental stewardship, and advocate for caring relations to the human and nonhuman world. Social media platforms encourage users to make content quickly, often by reposting and remixing. Combining the popular slogan #landback with Stewart's image demonstrates the way climate activists are drawing connections between Indigenous rights and climate justice. By using social media to generate poetic responses to one another, activists are able to bring multiple ideas and movements into conversation with one another.

Furthermore, thanks to new public-facing roles, youth activists' position as key figures can question societal assumptions. Seeing youth as leaders can reframe children and the educational discourses through which they are often defined.

In alignment with recent scholarship about the political agency of youth, the roles taken up by youth climate activists prompt a reconsideration of the dominant Western views of children. As a large-scale global movement, SSC effectively challenge obstructive, oppressive relations between children and adults undergirded by conceptions of children as nonpolitical actors, receivers of knowledge, and emblems of innocence. In many educational situations, these underlying assumptions are commonplace and inform adult/child

Figure 1.1. Pas Finii. This piece is titled *Pas Finii*, which means "not finished" in Michif and refers to the ongoing negotiation over Canadian economic resources like those listed on the map. The completed piece, titled *They Make a Well Beaten Path* uses this Métis style floral beadwork on this economic map from a 1947 Dent's Canadian School Atlas. In 2021, this piece was acquired in the collection of Indigenous Art Centre, Crown-Indigenous Relations and Northern Affairs Canada, Gatineau, Quebec.

Source: Kim, Gullion Stewart @kim.art4life. 2020. "Beadwork Is Coming Along on the 1950s Economic Map of Western Canada." *Instagram*, September 16, 2020. https://www.instagram.com/p/CFNR60FB1UE/

relations, pedagogy, curricula, and educational design. In climate activist circles, adults are either urged to act as political agents on behalf of children or to act to prepare youth to be future political agents. Yet, as evinced by the youth activists and elucidated in contemporary theorizations on the political agency of children, youth are already agents of change. Even at very young ages, children not only accept the roles presented to them by adults, but question, avert, and transform them. The aversion exemplified in SSC has transformative power. Through public action, youth activists highlight their social categorizations as both subjects and objects, capable of making change while still seen as needing to be told what to do. In so doing, youth are teaching adults and the wider public to rethink complex arrangements between the present and future generations.

In this chapter, I first explore the junctures between youth activism and social media. Following the view that people experience the world in relation to broader narratives that shape identities and perspectives, I argue that educational structures must account for this widespread global shift in youth/ climate narratives bolstered by communication through platforms such as Instagram, Twitter, and TikTok. Further, understanding the social and historical contexts in which youth operate is essential in the ongoing task of determining which narratives are being reproduced as a method of finding sites where new horizons are possible. I then share work by a youth activist spoken-word poet as a lead-in to an activity called the "100s" to examine social media representations of SSC. Akin to the modes of communication youth activists take part in on social media, the 100s encourages brevity, experimentation, and analysis while advancing the perception that youth activists are the principal authors to read for lessons toward climate justice— hashtags and posts the texts they write. Lastly, I argue that educators are uniquely positioned to advance climate justice as envisioned by SSC. SSC have gained traction as activists highlight their status as youth to call attention to the incongruities between climate uncertainties and educational systems designed around assumed and known futures. Educators already reckoning with increasing environmental precarities can act as essential allies in reimagining schooling by advancing justice-driven reconceptualizations of children and the climate crisis.

SOCIAL REMAKING AND SOCIAL MEDIA LITERACY

As John Dewey (1897) observed,[3] societies turn to new ways of understanding children in times of crisis. In the field of childhood studies, there is a burgeoning canon of work articulating the reconceptualization of children taking place in this new ecological epoch, the Anthropocene. Childhood studies scholars suggest that the climate crisis might offer openings to re-story notions of the child, dismantle educational systems that oppress based on the category of one's age,; and reshape educational spaces to honor children and their needs.[4] Dominant Western worldviews often position children as subservient to adults and, in turn, treat them as such in social and political terms. In the United States, children and adolescents (legally defined as "minors" until the age of eighteen) are expected to have less economic, political, and social agency. In this worldview children are considered to be less intellectually and emotionally developed, lacking "reason" which has long been considered necessary to participate in adult society. These are the tenets upon which US educational systems are often designed, leading to top-down hierarchies and pedagogical practices wherein knowledge is presumed to be gained through a unidirectional transference from teacher to student, rather than shared, horizontal structures to support the reciprocal nature of learning. Importantly, toward this end, the youth climate movement is resetting adult/child hierarchies, repositioning youth and adults as equally important contributors in efforts toward social and environmental justice. To many, "childhood" and "political agency" are antithetical. Therefore, as ethnic studies professor David Alberto Quijada (2009) posits,[5] when youth take political action, they offer a way of rethinking youth within the youths' own paradigm. Social media technologies have contributed to this reorientation, positioning youth as public content creators who can reciprocally engage in global and political dialogues. Contemporary youth activists demonstrate that they are stakeholders ready to make valuable contributions, reconfiguring long-held power relations.[6]

A paradigmatic example of a powerful youth can be seen in celebrity activist Greta Thunberg. Thunberg began her strike with a sign that read "Skolstrejk för klimatet" (School Strike for Climate). This action garnered attention as youth across the globe joined in to skip school, making headlines by refusing to participate in the primary social institution (besides family) meant to prepare them for their adult futures. Thunberg used social media to encourage other young people to become leaders, rather than followers, inspiring a diverse movement as youth expressed their own visions of climate justice relevant to their own communities.

From Thunberg's calls to action, Vanessa Nakate began to protest outside the parliamentary gates in Kampala, Uganda. Nakate used social media to encourage others to join, eventually starting a local strike large enough to gain international attention. She founded Rise Up to amplify Africans' voices and bring focus to the Congo Basin. In Wisconsin, USA, while organizing local SSC protests, Max Prestigiacomo and Stephanie Salgado founded Youth Climate Action Team (YCAT). Prestigiacomo and Salgado claimed that Instagram has been their sole organizing platform saying, "just a few clicks on social media can change hearts and minds." They have since become local leaders advocating for a diverse range of issues. As of 2020, Prestigiacomo had been elected to serve as alderman on the Madison City Council, becoming the youngest elected official in the United States, and Salgado had been given a seat on the Governor's Task Force on Climate Change. During a fall 2019 climate strike a reporter asked Salgado when the next march would take place, to which she answered, "when Greta Thunberg decides to make it."[7]

Tapping into the emergent youth climate movement, SSC have become central in contemporary climate narratives. Yet, movements do not happen in a vacuum; they exist within the sociopolitical landscapes of their time. These strikes have been influenced by predecessor Western social movements, including the environmental and labor movements of the twentieth and twenty-first centuries. For example, Thunberg's group Fridays for the Future uses the cartographic globe as their organizational logo. Since NASA's 1968 *Earthrise* photo of Earth from space, environmentalists have latched onto the globe to symbolize the interdependence of life on this planet. Fridays for the Future's logo provides a direct reference to previous generations of environmentalism.

However, one key aspect that distinguishes contemporary social movements from past movements is the use of social media, which has amplified the interconnectedness of social movements. Recent research on youth and social movements has shown the importance of the internet as a defining site for political engagement.[8] Social media has become essential in forming and introducing new political relationships and ideological stances. Thus, for SSC, the rise of momentum-based organizing work, hashtags such as #NoDAPL and #BlackLivesMatter, have proved formative.

For a prominent example of social media use in activism, consider the protests in Standing Rock Reservation to halt the construction of the Dakota Access Pipeline, which began when youth from Standing Rock and surrounding Native American communities launched a social media campaign, ReZpect our Water. In a viral video, Tokata Iron Eyes, AnnaLee Yellow Hammer, Precious Winter Roze Bernie, and Winona Gayton call on intergenerational and international participation to resist the pipeline and defend Indigenous sovereignty. The hashtag #NoDAPL became a shared slogan signifying entanglement between the genocidal and colonizing foundation of US expansion and threats posed to clean water and a livable planet. These issues have stayed front and center in continued iterations of youth climate movements across the globe. In addition, the mobilizing hashtag and organization #BlackLivesMatter, created by Alicia Garza, Patrisse Cullors, and Opal Tometi, has clarified the interconnected nature of racial injustice and economic exploitation and the correlation of a violent military system abroad and a militarized police practice at home.[9] Following the shooting at Marjory

Stoneman Douglas High School in Parkland, Florida, youth activists used Twitter as a primary venue to organize March for Our Lives.[10] Organizers spoke of their debt to #BlackLivesMatter as young Black activists across the United States joined in solidarity with the march to end gun violence. It is March for Our Lives that Thunberg cites as her inspiration to begin her School Strike for Climate.

In recent years, Instagram has become a primary place for young activists to absorb and communicate stories of climate change. Through videos, images, and texts Instagram allows users to share succinct multimodal stories, which are woven into narratives through a networked structure that users navigate through feeds, likes, and hashtags. In September 2019, in an interview on a local radio station, leaders from the Wisconsin-based Youth Climate Action Team urged adults to rethink their prejudices toward Instagram, suggesting that youth "scrolling throughout the day" are being exposed to the climate crisis. They claimed Instagram had been their most powerful organizing platform, speculating that "just a few clicks on social media can change hearts and minds." While these comments highlight the influence of Instagram, they ignore valid concerns regarding the ways social media creates "risky settings" for youth to engage political dialogue,[11] exposing youth to increased misunderstandings and vulnerabilities.

Instagram is a photo and video-sharing service owned by Facebook that relies on user-generated content shared as people connect with others on the site. These connections help generate links between disparate data sets based on social interaction. However, the design of Instagram stands in contradiction to the aims of climate justice in that it is predicated on extraction and commodification and encourages screen time, which can result in users identifying with causes without taking action. The core of the commercial value and activity is premised on captivating attention and instrumentalizing social interaction so online services can collect, process, and synthesize data to profile users and generate targeted advertising and personally relevant suggestions.[12] Many authors have raised concerns about these practices regarding issues of surveillance and privacy[13] and the way in which the algorithms that dictate our lives are opaque, unregulated, and reinforce discrimination.[14] According to design critic Alexandra Lange, Instagram's form favors bold colors and patterns and few elements and details. This glossy, reductive, and universalist aesthetic fuels the "Instagram gaze" of expensive lifestyles, which can create anxiety for people who look and live differently. Regardless of these drawbacks, as researchers have demonstrated, youth spend a considerable amount of time and effort with social network sites.[15] Thus, to address these concerns, educators cannot avoid social media. Supporting media literacy can offer space for youth to explore their identities[16] and respond to limiting mainstream narratives.[17] To empower students to engage proactivity with social media, educators need to foster a critical literacy, systemic thinking, and an awareness of user privacy rights and the impacts media consumption has on their perceptions and behaviors.

Precisely because young people use social media to read and write their world, educators can cultivate media literacy from their practices to promote climate justice. As posited by Paulo Freire, reading the word is a continuous process of reading the world. Reading cannot occur independently from the world of the reader. Now, readers are enmeshed in new media technologies that are rapidly changing how they read and write their worlds. Youth are active in participatory cultures, to borrow from Jenkins,[18] involved in both the production and consumption of media that can be widely and instantaneously shared. Youth voices, previously left in the backyard or on the playground, are now being amplified by social media platforms as part of a larger public discourse. Correspondingly, what counts as literature, and by extension, writing, has expanded beyond print-based texts to include artifacts found in video games, film production, and social media.[19] The social media of social movements offer educators new reading materials as well as new compositional forms. This literature can drive classroom inquiry to enable students to reflect more critically on climate change and thus write a more livable future.

XIUHTEZCATL MARTINEZ: SPOKEN WORD POET AND YOUTH CLIMATE ACTIVIST

Youth activist Xiuhtezcatl Martinez has been in the spotlight since childhood. At six years old Martinez lamented at an environmental event how "most kids don't even know the world *is* sacred." By 2019, at age nineteen, Martinez's Instagram had over 100,000 followers. He uses social media as a primary venue to publish and circulate his beliefs about the nonhuman world. This section acknowledges Martinez's posts as climate literature to reshape assumptions about what "counts" as literary text in alignment with expanding definitions of literacy education beyond print text, as discussed in the previous section. The following paragraphs frame the work of Martinez as an example of an influential youth climate activist author while offering pedagogical guidance on textual analysis.

In alignment with Martinez's Indigenous culture, his posts and practices embrace the oral tradition of sharing knowledge through songs, prayers, and music. As Indigenous peoples increasingly demand their oral traditions be taken more seriously in educational spaces,[20] including authors like Martinez offers an opportunity to affirm non-Western cultures. Further, oral texts can prompt educators to shift literary analysis away from its roots in dominant Western practices toward more generative methods. Indigenous authors have articulated that oral texts are best understood by engaging with messages presented to the listener repeatedly in a dialogic space where conflicting interpretations are welcomed.[21] This process can produce new and diverse answers from students treated not as empty vessels, but as cultural beings capable of creating valid responses. In light of the impossible questions posed by the climate crisis, the topic necessitates teaching outside of "right" or "wrong" to engender holistic understandings animated by the thoughts and feelings that arise.

Below is the transcribed text from a video of Martinez reciting a piece of spoken word posted to his Instagram feed April 2020. As a prose poet, Martinez communicates in open-ended ways, passing along knowledge that nature is not separate from humans, about power and resistance, indigenous lifeways, and that climate change encompasses issues of politics and justice. He regularly demonstrates intertwined commitments to the Movement for Black Lives, Indigenous sovereignty, and environmentalism. Throughout Martinez's posts, his poetry pushes the boundaries between mediums and genres to offer a holistic view of climate action as boundless and imaginative, a product of civics, art, protest, and performance. I encourage teachers to engage with this type of material by performing recitations or sharing video rather than using written text. The goal in listening or reading is not to agree or disagree with Martinez, but to engage responsively, asking: What do you think and how do you feel?

> We build a legacy and leave it and y'all can't say my name, but you know it's me when you see it. I am indigenous, product of genocide, colonization stole my past, but I refuse to hide. Never alone though we hungry like a pack of wolves, howling at the moon, indigo is runnin through the woods. Tired of bureaucracy, sick of broken democracy, stories of the end fulfilling sinister prophesies. We overthrowin kingdoms, ready to break free from, leaders that lead with barrels of oil that be leaking, and leadership's a beacon. We leadin' ships to freedom.[22]

To elicit poetic responses to poetry, questions might include: what resonates? What comes to mind? What flickers? What memories are evoked? What colors or images do you see? What connotations are you drawing? This can look like words or sentences, as fragmented and nonlinear as makes sense to the person responding. This chapter makes the argument that the literature published on social media by youth activists can offer educational settings the chance to engage with new forms of composition that illuminate the intricacies of the climate crisis. In the next section, I provide a lesson plan to engage with social media texts like that of Martinez to generate new poetic responses that can be shared on social media as well as through in-class recitations and oral performance.

THE 100S: A PEDAGOGICAL PRACTICE

To integrate the lessons of the SSC into more formal educational contexts, I want to suggest a practice called the 100s. As developed by English professor Emily Bernard, the 100s comprise a writing activity predicated on the notion that no piece can go over 100 words. With a strict word count and focus on brevity, this writing practice aligns with current youth reading and writing practices on social media. In Bernard's open-ended approach, a writer launches their 100-word piece from anything that resonates or sparks their interest. This unassuming, flexible concept can be applied across numerous settings from classrooms to writing groups. In the "100-Word Collective," a collection of published 100s, anthropologist Circe Sturm posits that this practice is enticing because the abbreviated format encourages greater dialogue and because writers of the genre tend to feel free to experiment with form and voice.

I learned of this practice while attending a spoken-word event of Kim TallBear, an esteemed professor of Native studies at the University of Alberta. TallBear has since applied this practice as a teacher, instructing her students to choose their favorite familiar slogan or hashtag and write 100 words. In this way, the activity mirrors the plurality of perspectives across the decentralized format of social media. For some the word count is flexible, but for TallBear each piece is exactly 100 words. TallBear has cited a more rigid adherence to word count as a helpful constraint in a discipline that she finds generative and inspiring.[23] TallBear's students use the 100s to describe Instagram posts as well as to create Instagram posts. With no "right way" to interpret a post, TallBear's 100s provide her and her students a framework within which they can more freely read, write, and interpret the world.

The 100s offer a means for anyone to engage dialogically with the texts produced on social media: a parent and child, a classroom community, or a group of friends. As such this activity does not serve existing hierarchical structures and can be easily adapted to numerous sites and situations. In a recent workshop for Earth Day on visual idioms of the environmental movement, I used the 100s with thirty participants. Before dividing attendees into breakout groups of three, I chose ten prominent symbols such as the globe and the recycling logo. The groups were assigned a symbol and provided a list of guiding questions to discuss the rhetorical underpinnings of the symbol. Then participants were encouraged to find examples of these symbols being deployed on the social media accounts of youth activists. At the end of the session, each person was tasked to choose one social media post to prompt a 100-word piece. I later distributed the 100s to the group, each response below the corresponding chosen symbols. By focusing on existing idioms from environmental visual rhetoric, this activity invited adults into the sophisticated nature of youth climate activism, showing a nuanced perspective of public idealizations of environmentalism. Undermining the notion that teachers have knowledge and students receive knowledge, this workshop models a more decentralized configuration for participates to learn from one another.

When considering the use of Instagram in a lesson plan, it is important to understand the context and risks of this social media service outlined previously in this chapter. It is also noteworthy that some students do not use social media, for reasons including issues of internet and device access.[24] Considering these factors, educators can ask students to participate as insiders or outsiders depending on their interests and expertise in social media and climate activism. Outside both climate activism and social media, educators and students can use this technique to examine key messages without any expectation of participation or posting. In this way, the 100s can accommodate educator allyship with students without explicit associations to the youth climate movement. On the other hand, for educators and students eager to forge unambiguous ties with climate activism on social media, the 100s can be assigned together with Instagram feeds to create brief, engaging texts that can be posted online and viewed as part of this ever-expanding dialogue.

Personally, I advocate for giving students agency over their own social media posting and profiles. I do not ask students to actually generate posts, just to examine the posts. By doing this, I focus on fostering the ability to navigate social media feeds that are flooded with visual rhetoric, advertisements, and subject to "micro-targeting" of political messages. For students on and off social media, I believe the 100s can help to animate a desire to write their own stories—to not feel like their lives are determined by narratives they had no part in composing. To accomplish this, there are many alternative empowering ways for people to share their voices during a 100s lesson plan, ranging from the creation of group zines to spoken-word performances.

THE 100S: AN ANALYTICAL CREATION

One of the fluidities of this practice is that it can serve both pedagogical and analytical aims.

Below is a selection of ten 100s I've written prompted by slogans and hashtags from SSC. I have adopted the practice as an analytical exploration, a method to reflect upon, and respond to, activist-generated texts I routinely encountered during a case study I conducted as a doctoral student in education and environmental studies. Over the past few years, I attended strikes, panels, meetings, and public events and followed media representations through social media, news, radio, speeches, and political blogs. Instagram became a primary tool for me to engage with many youth activists' perspectives and strategies. Thus, to examine how representations of the youth climate movement replay, remix, and transform climate narratives in ways that also shift narratives about adult-youth relations, I analyzed 600 Instagram posts made between August 20, 2018, and December 2, 2020. Phrases repeatedly appeared on Instagram feeds such as "The earth or nowhere," "No one is too small to make a difference," and "I'm sure the dinosaurs thought they had time too." The youth activists' use of Instagram inspired me to explore this writing style rooted in reciprocity. Following TallBear, I offer my 100s as an analytical, poetic, and shareable practice to expand upon significant hashtags and slogans that merit continued attention. Each 100 is titled with the corresponding hashtag or slogan I encountered during my research, and the title is not included in the word count.

Why Study for a Future That No Longer Exists?

Adults desire to protect children (from adulthood) or prepare children (for adulthood). But the secret is out. The future will not look the same as the past. Adults, having failed to protect youth from climate change, still implore, go to school and you'll become productive members of society. Yet, maintaining this future relies on imaginary pasts that displace untidy or disordered histories. Configurations of the child, an emblem of innocence distinctly envisioned as white, continue to partition resources to maintain white futures—futures that are now at stake. A faulty premise, a myth exposed. Time to break with the façade.

Stop Shitting on My Future

In August 2018, Greta Thunberg decided to stop going to school. Instead, she sat quietly on the cobblestones outside of Stockholm's parliament handing out leaflets that read, "stop shitting on my future." This slogan is honest and straightforward, hallmarks of Thunberg's pleas and her unapologetic way of confronting adults. Thunberg's commitment to the status quo is at the core of her persuasion. She'd rather be in school, would rather a future of careers, comfort, and predictable social order. Her power lies in her ability to articulate that this future is what is under threat as the climate catastrophes barrel on.

Denial Is Not a Policy

Climate denial is a deficit in shared practice. Policy lays out a course of action. As people became aware of the finitude of the planet's recourses, corporations and "elites" did all they could to maintain the status quo and preserve their power. Responsibility redirected to the consumer. Attention diverted away from the earth and its inhabitants. Distraction and delusion obstructive cornerstones of strategy. Radically veering us off course. To reroute, educators can center how to live in the same world, share the same culture, face up to the same stakes, and perceive a landscape that can be explored in concert.

If You Don't Act Like Adults, We Will

Youth are drawing attention to their role as children. "Child" vulnerability lies in opposition to "adult" power. If adults are presumed reasonable and mature, children are reckless and emotional. But we'd lose so much if children acted like adults; we'd lose their risk-taking, creativity, emotional connectedness. Child strengths now radiate as they take up space and positions of public leadership. In taking political power, youth activists blur adult/child boundaries. By simultaneously reifying and dismantling what it means to be a child, youth activists lay new roots for emancipatory climate justice. Can this change what it means to be an adult?

This Can't Wait Till I'm Big

Sianne Ngai asks, is "cute" a way to aestheticize powerlessness? I wondered this as I watched adorable toddlers holding up signs, clunky cardboard too big for them to carry. The cute object is most fully itself when weak and in need of protection, often politicized through propaganda regimes—the helpless starving child, the vulnerable polar bear cub, and the clumsy baby penguin. By holding these self-referential signs, youth are able to render cute as capable. Instead of being infantilized (drawing on conceptions of infants and baby animals as incapable beings), with this phrase they brandish their "cuteness" to demand action.

Our House Is on Fire

A simple metaphor energizing urgency. This is an issue for the here and now. Greta doesn't want hope, she wants panic. Can you feel the fear she feels every day? Record heat waves, forest fires, and protests over mines and pipelines garner national attention. There is no time to wait. The flammable materials that power homes, vehicles, and factories seal our fate. Governments still subsidize fossil fuels rather than determine plans for prevention. This is a betrayal. It is not only that our house is on fire, but we lack a fire department. If earth is home, let's put out the blaze.

#Thereisnoplanetb

US environmentalist yearnings are hitched to the frontier myth—an empty place to escape, develop, colonize. NASA sees space as the next phase of US evolution: "the next step of growing up is going up." But, oh. Oops. There is no safety valve. The extraterrestrial frontier is predicated on falsities: the West was not empty (the genocide of Indigenous populations was an essential component of westward expansion) and space is not empty (it is actually full of space garbage). Instead of inventing an out-of-this-world reality, youth activists opt for the messy trouble of forging alliances toward an earthbound future.

#TMTShutdown

Come down from space. Inhabit the Earth. Since July 2019 protesters have resisted the proposed construction of a Thirty Meter Telescope (TMT) atop Mauna Kea, the volcanic peak where kia'i Hawaiians practice daily rituals. Astronomers say a TMT will offer insight into the earliest days of the universe. A universe, seen from nowhere, stretches from our bodies to the most distant galaxies. But activists fear TMT will further desecrate the summit. I see a new and nuanced demand for scientific agendas that respect the conjoined human and nonhuman agents that constitute Mauna Kea. Partake in the creation of earthly knowledge.

Activism Is Learning

New ways of relating to climate change are reshaping how and where learning takes place. Australian Prime Minister Scott Morrison beseeched, "less activism and more learning in schools." But too many people saw through his veiled attempt to undermine requisite transformation. Activism offers students a means to teach and to learn. Now, protesters are learning what counts as action, adult nostalgia for past progress in signs, slogans, marches meet the moment of hashtags and memes. An activated movement learns of its power, as the protesters demonstrate a shift in pedagogical authority from parents and teachers toward the body politic itself.

ALLYSHIP AS/IN EDUCATION

Like social media, school systems organize the lives of many. Schooling can reproduce injustices as it segregates young people, surveils them, and often dictates (and erases) histories and worldviews. In both classrooms and curricula, youth voices are too rarely privileged. Yet, schools can also house meaningful, empowering exchanges of care and knowledge. Toward this end, SSC provides society clear critiques of school's inadequacies. As youth climate activists have taught us, it's time for a global reckoning that the future will not be the same as the past, and education will only remain relevant insomuch as it can attend to the environmental crises that disproportionately threaten young people, particularly those already marginalized due to economic, racial, and social inequities. Student strikers demand to be heard, sharing stories of newly flourishing youth/climate narratives. While students strike, forests burn, and pandemics sweep the globe, educators can either hang onto an antiquated system or aim for relevance best achieved through allyship.

A fundamental aspect of allyship is the reshaping of power dynamics. The 100s, as a pedagogical and analytical practice, can operate on multiple levels to begin to privilege youth voices, reshape teacher/student relations, and promote more in-depth engagement with youth activists' teachings on climate justice. Mirroring the decentralized nature of activism on social media, the 100s can subvert hierarchy by harnessing educators' positionality to center youth voices. On Instagram, a uniform structure and lack of gatekeeping encourages the proliferation of diverse content. Yet, the company consolidates control over how and where information can travel. Counter to Instagram, the 100s do not deny the uneven power dynamics in classroom settings but set the parameters to work with youth, enacting what Vivian Chávez and Elisabeth Soep call a "pedagogy of collegiality."[25] Rather than romanticizing youth leadership as something done in the absence of adult relations, this pedagogy of collegiality emphasizes youth voices without abandoning them to do the work entirely on their own.[26]

Further, by analyzing the primary texts of youth climate activists, educators can support cultural shifts that recognize climate change as entangled with issues of justice and reconceptualize children as vital leaders in efforts to address humanity's most pressing concerns. The confluence of smartphones and social media has led to new technologies that activists use to educate and organize around issues of police brutality, settler colonial violence, and global climate catastrophes. In response, educational structures must integrate the teachings of youth activists advocating for more nuanced, intersectional visions of justice. Inviting students to engage critically with SSC gives value to youth activists' contributions and allows for multiple interpretations. The climate crisis requires a reorientation, a commitment to join with those who are imagining and constructing worlds that deserve a future, worlds committed to diversity and taking care of the marginalized human and nonhumankind. As Donna Haraway argues, rather than a distant goal, ongoing futures rely on us building thick, robust presents with each other, on working and playing with each other in ways that make sense now. Teachers no longer holding on to a predetermined future for which they must prepare their students can instead create and facilitate structures for students to imagine and experience a more just and livable world. The 100s offered in this chapter, as activity and examination, are one small suggestion embedded in this broader call to rethink both education and climate change by taking seriously the words and worlds of youth.

NOTES

1. Rob Nixon, "The Anthropocene: Promise and Pitfalls of an Epochal Idea," *Edge Effects,* November 6, 2014. http://edgeeffects. net/anthropocene-promise-and-pitfalls

2. Finis Dunaway, "Gas Masks, Pogo, and the Ecological Indian: Earth Day and the Visual Politics of American Environmentalism," *American Quarterly, 60,* no.1, (2008): 67–99.

3. John Dewey, "My Pedagogic Creed," *School Journal,* 54, no. 8, (1897): 77–80.

4. Alexandra Lakind and Chessa Adsit-Morris, "Future Child: Pedagogy and the Post-Anthropocene." *Journal of Childhood Studies* 43, no.1 (2016): 30–43; Veronica Pacini-Ketchabaw and Fikile Nxumalo, "Regenerating Research Partnerships in Early Childhood Education: A Non-Idealized Version," in *Research Partnerships in Early Childhood Education: Teachers and Researchers in Collaboration,* edited by Judith Duncan and Lindsey Conner, 11–26. New York: Palgrave Macmillan, 2013; Rebekah Sheldon, *The Child to Come: Life after Human Catastrophe* (Minneapolis, MN: University of Minnesota Press, 2016).

5. David Alberto Quijada, "Youth Debriefing Diversity Workshops: Conversational Contexts That Forge Intercultural Alliances across Differences," *International Journal of Qualitative Studies in Education* 22, no. 2 (2009): 449–68.

6. Jacqueline Kennelly, "Citizen Youth: Culture, Activism, and Agency in a Neoliberal Era," in *Contemporary Youth Activism: Advancing Social Justice in the United States,* edited by Jerusha Conner and Sonia Rosen, 79–92 (Westport, CT: Praeger Press/ABC-CLIO Books, 2016).

7.Sam Jones and Tenzin Woser. "Climate Strike Draws Hundreds to Downtown Madison," *Madison Commons,* September 25, 2019. https://madisoncommons.org /2019/09/25/climate-strike-draws-hundreds-to-downtown-madison/

8. Melvin Delgado and Lee Staples, *Youth-Led Community Organizing: Theory and Action.* (New York: Oxford University Press, 2008); Ben Kirshner, "Guided Participation in Three Youth Activism Organizations: Facilitation, Apprenticeship, and Joint Work," *Journal of the Learning Sciences* 17, no. 1, (2008): 60–101; Dana L. Mitra, "Adults Advising Youth: Leading While Getting out of the Way," *Educational Administration Quarterly* 41, no. 3 (2005): 520–53.

9. Bernadine Dohrn and Bill Ayers, "Young, Gifted, and Black: Black Lives Matter!" in *Contemporary Youth Activism: Advancing Social Justice in the United*

States, edited by Jerusha Conner & Sonia Rosen, 79–92. (Westport, CT: Praeger Press/ABC-CLIO Books, 2016).

10. Carrie James and Megan Cotnam-Kappel, "Doubtful Dialogue: How Youth Navigate the Draw (and Drawbacks) of Online Political Dialogue," *Learning, Media and Technology* 45, no. 2 (2019): 129–50.

11. Kjerstin Thorson, "Facing an Uncertain Reception: Young Citizens and Political Interaction on Facebook," *Information, Communication & Society* 17, no. 2 (2014): 203–16; Ariadne Vromen, Brian D. Loader, Michael A. Xenos, and Francesco Bailo, "Everyday Making Through Facebook Engagement: Young Citizens' Political Interactions in Australia, the United Kingdom and the United States," *Political Studies* 64, no. 3 (2016): 513–33.

12. Amy Gaeta,. "Do Algorithms Know Your Body Better Than You?" *One Zero.* October 27, 2019. tps://onezero.medium.com/do-algorithms-know-your-body-better-than-you-2f1c7d241144; Tim Wu, *The Attention Merchants: The Epic Scramble to Get inside Our Heads* (New York: Knopf, 2016).

13. Shoshna Zuboff, *The Age of Surveillance Capitalism: The Fight for a Human Future at the New Frontier of Power* (New York: Public Affairs, 2019); David Lyon, *Surveillance as Social Sorting: Privacy, Risk, and Digital Discrimination* (New York: Routledge, 2003).

14. Cathy O'Neil, *Weapons of Math Destruction: How Big Data Increases Inequality and Threatens Democracy* (New York: Crown Publishers, 2016).

15. danah boyd, "Why Youth (Heart) Social Network Sites: The Role of Networked Publics in Teenage Social Life," in *Youth, Identity, and Digital Media*, edited by David Buckingham, 119–42 (Cambridge, MA: The MIT Press, 2007); boyd, danah and Nicole Ellison, "Social Network Sites: Definition, History, and Scholarship." *Journal of Computer-Mediated Communication* 13, no. 1 (October 2007): 210–30; Christine Greenhow, "Youth Learning and Social Media," *Journal of Educational Computing Research* 45, no. 2, (2011): 139–46.

16. Erica Rosenfeld Halverson, "Film as Identity Exploration: A Multimodal Analysis of Youth-Produced Films," *Teachers College Record* 112, no. 9 (2010): 2352–78.

17. Jen Scott Curwood and Damiana Gibbons, "'Just Like I Have Felt': Multimodal Counternarratives in Youth-Produced Digital Media," *International Journal of Learning and Media* 1, no. 4 (2009): 59–77; Jeff Duncan-Andrade, "Urban Youth and the Counter-Narration of Inequality," *Transforming Anthropology* 15, no. 1 (2008): 26–37.

18. Henry Jenkins, *Convergence Culture: Where Old and New Media Collide* (New York: New York University Press, 2006).

19. Erica Rosenfeld Halverson and Damiana Gibbons,. "'Key moments' as Pedagogical Windows into the Digital Video Production Process," *Journal of Computing in Teacher Education* 26. no. 2 (2010): 69–74; Mizuko Itō, *Hanging Out, Messing Around, and Geeking Out: Kids Living and Learning with New Media* (Cambridge, MA: MIT Press, 2010).

20. Esther Ilutkik, "Eleven Years of Implementing Traditional Yup'ik Oral Stories in the Elementary Classroom." Paper presented at Orality in the 21st Century: Inuit

Discourse and Practices. Proceedings of the 15th Inuit Studies Conference, Paris: INALCO, 2009. http://www.inuitoralityconference.com.

21. Julie Cruikshank, "Oral Tradition and Oral History: Reviewing Some Issues." *The Canadian Historical Review* 75, no. 3 (1994): 403–18. Project MUSEmuse.jhu.edu/article/574633

22. Xiuhtezcatl Martinez (@xiuhtezcatl), "My Heart Is Heavy. @berniesanders Thank You for Your Sacrifice, Your Courage, Your Leadership," Instagram, April, 8, 2020, https://www.instagram.com/p/B-ulHQQg2bu/

23. Kim TallBear, "Critical Poly 100s," *The Critical Polyamorist,* July 14, 2020. http://www.criticalpolyamorist.com/critical-poly-100s.html.

24. Katy Pearce and E. Ronald Rice, "Digital Divides from Access to Activities: Comparing Mobile and Personal Computer Internet Users," *Journal of Communication* 63, no. 4 (2013): 721–44.

25. Vivien Chávez and Elisabeth Soep, "Youth Radio and the Pedagogy of Collegiality," *Harvard Educational Review* 75, no. 4 (2005): 409–34.

26. Barbara Ferman and Natalia Smirnov, "Shifting Stereotypes and Storylines: The Personal and Political Impact of Youth Media," in *Contemporary Youth Activism: Advancing Social Justice in the United States*, edited by Jerusha Conner and Sonia Rosen, 185–201 (Westport, CT: Praeger Press/ABC-CLIO Books, 2015).

BIBLIOGRAPHY

boyd, danah. "Why Youth (Heart) Social Network Sites: The Role of Networked Publics in Teenage Social Life." In *Youth, Identity, and Digital Media*, edited by David Buckingham, 119–142. Cambridge, MA: The MIT Press, 2007

boyd, danah, and Nicole Ellison. "Social Network Sites: Definition, History, and Scholarship." *Journal of Computer-Mediated Communication* 13, no. 1 (October 2007): 210–30.

Chávez, Vivian, and Elisabeth Soep. "Youth Radio and the Pedagogy of Collegiality." *Harvard Educational Review* 75, no. 4 (2005): 409–34.

Cruikshank, Julie. "Oral Tradition and Oral History: Reviewing Some Issues." *The Canadian Historical Review* 75, no. 3 (1994): 403–18. Project MUSEmuse.jhu.edu/article/574633

Curwood, Jen Scott, and Damiana Gibbons. "Just Like I Have Felt": Multimodal Counternarratives in Youth-Produced Digital Media." *International Journal of Learning and Media* 1, no. 4 (2009): 59–77.

Delgado, Melvin, and Lee Staples. *Youth-Led Community Organizing: Theory and Action*. New York: Oxford University Press, 2008.

Dewey, John. "My Pedagogic Creed." *School Journal* 54, no. 8 (1897): 77–80.

Dohrn, Bernadine, and Bill Ayers. "Young, Gifted, and Black: Black Lives Matter!" In *Contemporary Youth Activism: Advancing Social Justice in the United States,* edited by Jerusha Conner and Sonia Rosen, 79–92. Westport, CT: Praeger Press/ABC-CLIO Books, 2016.

Dunaway, Finis. "Gas Masks, Pogo, and the Ecological Indian: Earth Day and the Visual Politics of American Environmentalism." *American Quarterly* 60, no.1 (2008): 67–99.

Duncan-Andrade, Jeff. "Urban Youth and the Counter-Narration of Inequality." *Transforming Anthropology* 15, no. 1 (2008): 26–37.

Ferman, Barbara, and Natalia Smirnov. "Shifting Stereotypes and Storylines: The Personal and Political Impact of Youth Media." In *Contemporary Youth Activism: Advancing Social Justice in the United States*, edited by Jerusha Conner and Sonia Rosen, 185–201. Westport, CT: Praeger Press/ABC-CLIO Books, 2015.

Gaeta, Amy. "Do Algorithms Know Your Body Better Than You?" *One Zero*. October 27, 2019. https://onezero.medium.com/do-algorithms-know-your-body-better-than-you-2f1c7d241144

Greenhow, Christine. "Youth Learning and Social Media." *Journal of Educational Computing Research* 45, no. 2 (2011): 139–46.

Halverson, Erica Rosenfeld. "Film as Identity Exploration: A Multimodal Analysis of Youth-Produced Films." *Teachers College Record 112*, no. 9 (2010): 2352–78.

Halverson Erica Rosenfeld, and Damiana Gibbons. "'Key Moments' as Pedagogical Windows into the Digital Video Production Process." *Journal of Computing in Teacher Education* 26, no. 2 (2010): 69–74.

Ilutkik Esther. "Eleven Years of Implementing Traditional Yup'ik Oral Stories in the Elementary Classroom." Paper presented at Orality in the 21st Century: Inuit Discourse and Practices. Proceedings of the 15th Inuit Studies Conference, Paris: INALCO, 2009. http://www.inuitoralityconference.com.

Itō, Mizuko. (2010). *Hanging Out, Messing Around, and Geeking Out: Kids Living and Learning with New Media*. Cambridge, MA: MIT Press.

James, Carrie, and Megan Cotnam-Kappel. "Doubtful Dialogue: How Youth Navigate the Draw (and Drawbacks) of Online Political Dialogue." *Learning, Media and Technology* 45, no. 2 (2019): 129–50.

Jenkins, Henry. *Convergence Culture: Where Old and New Media Collide*. New York: New York University Press, 2006.

Jones, Sam, and Tenzin Woser. "Climate Strike Draws Hundreds to Downtown Madison." *Madison Commons*, September 25, 2019. https://madisoncommons.org /2019/09/25/climate-strike-draws-hundreds-to-downtown-madison/

Kennelly, Jacqueline. "Citizen Youth: Culture, Activism, and Agency in a Neoliberal Era Palgrave Macmillan." In *Contemporary Youth Activism: Advancing Social Justice in the United States,* edited by Jerusha Conner & Sonia Rosen, 79–92. Westport, CT: Praeger Press/ABC-CLIO Books, 2016.

Kirshner, Ben. "Guided Participation in Three Youth Activism Organizations: Facilitation, Apprenticeship, and Joint Work." *Journal of the Learning Sciences* 17, no. 1 (2008): 60–101.

Lakind, Alexandra, and Chessa Adsit-Morris. (2018). "Future Child: Pedagogy and the Post-Anthropocene." *Journal of Childhood Studies* 43, no.1 (2016): 30–43.

Lyon, David. *Surveillance as Social Sorting: Privacy, Risk, and Digital Discrimination*. New York: Routledge, 2003.

Martinez, Xiuhtezcatl (@xiuhtezcatl). "My Heart Is Heavy. @berniesanders Thank You for Your Sacrifice, Your Courage, Your Leadership." Instagram, April, 8, 2020. https://www.instagram.com/p/B-ulHQQg2bu/

Mitra, Dana. L. "Adults Advising Youth: Leading While Getting out of the Way." *Educational Administration Quarterly* 41, no. 3 (2005): 520–53.

Nixon, Rob. "The Anthropocene: Promise and Pitfalls of an Epochal Idea." *Edge Effects,* November 6, 2014. http://edgeeffects. net/anthropocene-promise-and-pitfalls

O'Neil, Cathy. *Weapons of Math Destruction: How Big Data Increases Inequality and Threatens Democracy*. New York: Crown Publishers, 2016.

Pacini-Ketchabaw, Veronica, and Fikile Nxumaloile. "Regenerating Research Partnerships in Early Childhood Education: A Non-Idealized Version." In *Research Partnerships in Early Childhood Education: Teachers and Researchers in Collaboration*, edited by Judith Duncan and Lindsey Conner, 11–26. New York: Palgrave Macmillan, 2013.

Pearce, Katy, and E. Ronald Rice. "Digital Divides from Access to Activities: Comparing Mobile and Personal Computer Internet Users." *Journal of Communication* 63, no. 4 (2013): 721–44.

Quijada, David Alberto. "Youth Debriefing Diversity Workshops: Conversational Contexts That Forge Intercultural Alliances across Differences." *International Journal of Qualitative Studies in Education* 22, no. 2 (2009): 449–68.

Sheldon, Rebekah. *The Child to Come: Life after Human Catastrophe*. Minneapolis, MN: University of Minnesota Press, 2016.

TallBear, Kim. "Critical Poly 100s." *The Critical Polyamorist,* July 14, 2020. http://www.criticalpolyamorist.com/critical-poly-100s.html.

Thorson, Kjerstin. "Facing an Uncertain Reception: Young Citizens and Political Interaction on Facebook." *Information, Communication & Society* 17, no. 2 (2014): 203–16.

Vromen, Ariadne, Brian D. Loader, Michael A. Xenos, and Francesco Bailo. "Everyday Making Through Facebook Engagement: Young Citizens' Political Interactions in Australia, the United Kingdom and the United States." *Political Studies* 64, no. 3 (2016): 513–33.

Wu, Tim. *The Attention Merchants: The Epic Scramble to Get Inside Our Heads*. New York: Knopf, 2016.

Zuboff, Shoshana. *The Age of Surveillance Capitalism: The Fight for a Human Future at the New Frontier of Power*. New York: Public Affairs, 2019

Chapter Two

"But, What Difference Can I Make?"

Using Documentaries to Explore Environmental Advocacy in the Face of Climate Change

Carley Petersen Durden and Jared Durden

We believe general education classrooms should be transformative spaces in which students can address and resolve real-world problems in a nonthreatening atmosphere. As climate change is and will continue impacting human and natural species worldwide, we think it imperative to integrate climate-related discussion into general education curriculum. To that end, we recommend using documentary films as efficient, persuasive texts that engage students on an intellectual, ethical, and emotional level. Film is a medium in which students, as twenty-first-century learners, are well versed. In addition, film promotes interdisciplinary learning, mingling multiple subject areas, such as writing, rhetoric, science, and social issues, into a single course, perhaps even a single class session. Finally, film can help depoliticize climate-change threats, thereby inspiring students to focus on creative solutions rather than simply rehashing the difficulty of addressing problems.

A powerful form of storytelling, "[documentaries] depict real problems involving real people [that command] a deeper empathy [from] the audience . . . [and] . . . show a great potential to reach a wide range of [viewers]" in that they combine "visual, audial, and narrative experiences"[1] to which everyone can relate. Digital storytelling and documentary can inform effective pedagogy for the twenty-first-century classroom, allowing students to confidently explore complex social and environmental issues. Bernard

R. Robin suggests that digital storytelling succeeds for two reasons; first, students enter classrooms already utilizing multiple technologies as both creators and consumers; second, when combined with instructional guidance and/or written texts, visual media enhances comprehension and promotes key critical thinking and analytical skills.[2]

For several decades, educators in all disciplines have employed digital storytelling as an accessible and persuasive text that helps students identify, analyze, and prioritize specific social and environmental concerns. In writing courses, authors Veleda Boyd and Marilyn Robitalle use familiar television dramas to promote purposeful practice with idea extraction and concept synthesis.[3] Sociology instructor Jesse Daniels suggests that, when combined with instruction in media literacy, documentaries help students understand difficult course concepts and develop empathy for people of differing cultures and belief systems.[4] Alan S. Marcus and Jeremy D. Stoddard recognize that, when introduced to history classes, documentaries help students become responsible, democratic citizens with a healthy skepticism of agenda-driven rhetoric and imagery.[5] In evaluating a semester-long environmental studies course that pivoted documentary viewing to in-class discussion and guided reflection, Liu saw increased student appreciation for and interest in climate and environmental concerns that could also engender pro-environmental habits.[6]

Part of the appeal of documentary storytelling derives from affective rewards valued by mature, college-level students. "[E]motionally charged images, evoked through the context of adult learning, provide . . . a more profound access to the world by inviting a deeper understanding of ourselves in relationship with it."[7] Because documentaries use a combination of images, music, and strong, persuasive point of view, viewers are bound to become invested, often relating what they see on screen to their own life challenges. Showing the negative effects of climate change in images meant to shock or create fear is a popular, if problematic, way of raising awareness and provoking robust discussion. The intent is to create a sense of urgency and concern that will elicit adaptive behaviors for mitigating climate change that students can readily incorporate into their normal routines. However, some experts caution that fear-inducing representations of climate change, while widely employed in the public domain, may be counterproductive and recommend showing "non-threatening imagery and icons that link to individuals' everyday emotions and concerns"[8] as a useful and more appealing strategy. Documentaries that invoke viewer empathy invite oral, written, and online discussion and deepen awareness that lifestyle choices may exacerbate and/or mitigate environmental threats on a local, regional, and global scale. Valdez et al. confirm that students who engage in conversations about climate change with family and friends are more likely to adapt environmentally friendly attitudes and behaviors.[9]

While some students dismiss climate-change advocacy as politically or profit motivated and unworthy of serious debate, we find well-made documentaries provide talking points with which students can engage unfamiliar ideas and reexamine prior assumptions about their climate-challenged world. Although we are both general education instructors, we teach in two different disciplines, composition and physics, and therefore use two different documentaries in our classrooms to explore climate change issues.

Two inquiry-based approaches are presented in this chapter. The first is designed for a first-year writing classroom. The second is appropriate for an introductory physics course for nonscience majors. However, both can easily be adapted for secondary education/high school students. While focused on specific examples of documentary films, the guide that follows demonstrates adaptable strategies for engaging general education students in questions and conversations about climate change in a nonthreatening atmosphere.

BUGS IN THE WRITING CLASSROOM

Bugs follows a team of researchers and chefs on a global quest to discover how various cultures prepare and cook insects for human consumption. Motivated by the Food and Agricultural Organization of the United Nations' (FAO) grim projections on how population growth is affecting global food production, Josh Evans, Roberto Flore, and Ben Reade begin their journey believing insects may be the answer for a sustainable protein source for the twenty-first century. However, they soon discover how societal, political, and economic institutions are compromising global food production, accelerating environmental destruction, and exacerbating critical supply-chain issues. At a moment when environmentalists and business opportunists are promoting edible insects as sustainable and affordable alternative protein sources, Evans, Flore, and Reade begin to question the long-term effects of producing so-called green food using corporate agricultural models and methods.

In the writing classroom, *Bugs* can be used as content for a film review, a common assignment in first-year writing programs since reviews require writers to defend their recommendation for viewing the film with reliable evidence and analyze the filmmaker's use of rhetorical strategies. Before viewing the film, students should have some practice with analytical thinking and writing. A brainstorming session devoted to identifying the various components of film, such as plot, characterization, sound effects, and musical score, would provide useful vocabulary for assessing the film's storytelling and persuasive techniques. Supplementary reading on climate change's impact on biodiversity may be helpful as the film explores the impacts and challenges of the current model of global food production. The film could also be used as an introductory text for a semester or unit-long exploration of climate change and food production in a themed writing course.

When implementing a more comprehensive writing project, instructors might require students look at research supporting the decades-long decline of global insect populations—a concern not addressed in *Bugs* though recent studies confirm the world's insects are in dire trouble. As climate change research may be new to teachers themselves, they may invite students to treat them as members of their learning community when examining scientific articles such as Thomas et al.'s research, who attribute reduced butterfly numbers to an on-going destruction of the size and quality of butterfly habitat,[10] or Hallman et al.'s study, which states that Germany has experienced a 76 percent reduction in airborne insects in protected areas over the last 27 years.[11] Whatever the cause, Forister, Pelton, and Black emphasize "the severity of reported insect declines is . . . sufficient to warrant immediate action," recognizing that before definitive change can occur, the fate of insects must be seen by everyone as impacting species health across regional and global environments.[12] For example, Forister, Pelton, and Black suggest that "for variety, abundance and impact, insects have no rival among multicellular life on this planet," since they "connect innumerable other organisms in relationships that range from pollination to predation" in and across diverse ecosystems.[13] In addition, Biesmeijer et al. argue that "losses of pollinating insects [are] particularly troubling because . . . many agricultural and natural plant populations are dependent on pollination and . . . services provided by wild, unmanaged pollinator communities."[14] Thus, a candid, class discussion of the vital role insects play in environmental diversity and sustainability is both useful and appropriate for enhancing student learning.

Unfortunately, many Westerners take the view that insects are nuisances, pests, or vectors of disease. (The latter view came up during class when students explored insects as a viable food source but were concerned about the suspected origins of the COVID pandemic.) Thus, while they are likely aware of basic ecosystem functionality, students often have cultural biases that prevent them from caring about "lower-order" organisms, like "bugs." To that end, examining cognitive bias, a concept often explored in introductory writing and research courses, may provide another lens for students to examine any difficulty they have finding insects a species worthy of human attention and protection, not to mention a viable food source.

Teachers may find Kendra Cherry's ideas on raising bias awareness particularly helpful in a classroom setting where students are coming from diverse backgrounds and may count climate-change deniers among their friends and family, or who may be climate-change deniers themselves. Cherry's medically reviewed article,"What Is Cognitive Bias?," defines, explains, and contextualizes cognitive psychology in clear, understandable, prose accessible to student and other nonexpert audiences, so that students may feel comfortable sharing their views at home or work.[15] According to Cherry, and researchers like Dan Kanheman, cognitive bias "is a systematic error in thinking that occurs when people are processing and interpreting information in the world around them"—a noteworthy concern given that students are programmed to believe errors and mistakes are punishable offences.[16]

Several key takeaways from Cherry's article can deepen student understanding of the negative consequences accruing to cognitive bias while doing so in a productive, nonthreatening way. First, she discusses how the human brain evolved to use mental shortcuts when solving problems, based on prior knowledge and experience.[17] In fact, there is an evolutionary advantage to such quick thinking, particularly in moments of danger: humans make split-second decisions that promote their own safety and survival, which in turn supports the on-going survival of the species. Second, Cherry suggests that everyone is susceptible to bias, particularly experts.[18] Dan Ariely, a behavioral economist for example, posits that the capacity for bias may increase with one's level of expertise.[19] Moreover, while we may readily recognize bias in others, it's much harder to ferret out bias in ourselves. Too often we believe we "are objective, logical, and capable of taking in and evaluating all the information that is available" when a variety of unconscious factors inevitably colors our perceptions and actions.[20] Finally, and perhaps, most importantly, Cherry reminds that being aware of our own biases and not allowing them to unduly influence our decision-making will improve our empathy for, and interactions with, all those we encounter, whether human or otherwise.[21]

The film *Bugs* can be viewed in a single seventy-five-minute session or two sixty-minute sessions. However, it is recommended that the instructor pause the film to assess student understanding of and receptivity to insect-based cuisines. Students will likely have visceral reactions to the film, since it challenges their cultural expectations of what is considered food, and they will need to discuss their initial reactions before moving on to film analysis. Some students may express disgust at the thought of eating insects—a great moment to ask what are our common reactions to insects? Why don't Westerners see insects as a food source when so many other cultures do? How might these strong reactions suggest we are disconnected from nature as a culture? Given the statistics that open the film, why might it be important for Westerners to increase their knowledge of what is edible in the natural world?

After initial discussion, students are ready to begin analyzing the film. Prompts that remind students of the questions they will need to answer in a film review are helpful: how well does the film engage audiences? What rhetorical strategies does the film use? Which elements make the strongest contribution to the film, and why? Which elements fail to carry out the film's intention? Since these questions require analytical thinking, instructors should write questions on the board or projector screen, so students can write them down or give them some thought before responding.

Next, present clips from the movie, having students look for specific moments that use the elements of film to make a persuasive case for the audience. A recommended sequence of clips with discussion questions follows.

Opening Credits—2:50

The opening scene shows Evans and Reade preparing and serving sample gourmet airplane meals, using insects as ingredients, in an experimental food lab. This scene precedes opening credits accompanied by discordant music and foregrounding extreme close-ups of wriggling insects on a factory conveyor belt. Three stark statistics appear across the screen as the insects are packed in boxes and prepared for shipping: "By 2050 the world's population will be 9 billion people. The FAO of the UN estimates food production will need to increase by 70%. According to the FAO, insects will be a crucial part of our future diets."[22] The film's opening deliberately provokes discomfort in Western viewers, forcing them to confront common food challenges twenty-first-century citizens will face, and provides an opportune moment to begin analyzing the film's techniques while posing questions on how unsustainable agricultural practices impact biodiversity. Possible discussion questions may include: how are the statistics, close-ups, and music meant to make the viewer feel? Why does the film open this way? Students will recognize the connection between the film's techniques and their intended effect and likely reiterate their own discomfort when viewing the opening scenes. The statistics themselves can be interrogated: how do the opening statistics force the viewer to consider their role in unsustainable agricultural practices? While we may think about purchasing healthy, organic, or ethically sourced food, how much time does the average consumer spend considering the future availability of food?

Hunting for Grubs—5:56

Evans and Reade visit the Australian outback to hunt and cook prized grubs. After several failed attempts at finding the highly sought-after larvae, the researchers finally extract one from its casing and roast it in the ashes of a fire. Suspense builds as they pop roasted morsels into their mouths and begin chewing. Their faces and mouths fill the frame, sound fades, and a close-up captures a (locally) coveted dining experience. Evans and Reade make taste comparisons to a variety of familiar foods before delivering their surprising assessment: "It's delicious!"[23]

Problems with Industrial Diets—8:34

In Kenya, Evans and Reade learn about farming edible insects from Professor Monica Ayieko, who asks why two Westerners want her advice on edible insects. Reade explains; firstly, they are investigating the nutritional benefits of consuming insects on a routine basis; and secondly, as cultural anthropologists, they want to know why Westerners do not avail themselves of the 1,000-plus insect species that are safe for human consumption. Professor Ayieko shares that Africans themselves are returning to traditional diets that include insects to reduce the cancer risks, diabetes, and high blood pressure that come with eating processed food. Students should discuss how Western viewers may experience cognitive dissonance regarding their own dietary choices.

Alternatives to Industrial Diets—21:23

Evans and Reade visit a gourmet Mexican restaurant and meet the chef who serves escamole, a delicacy comprised of ant pupae and larvae. They then visit the farmers who care for the ants and actively work to preserve the local ecosystem. The farmers explain their cultivation methods and evaluate the risks and rewards of producing their own food in an environmentally sustainable way. In the next scene, Evans and Reade tour a polycultural garden in Uganda and indulge in honey produced by African stingless bees. Evans and Reade critique the FAO's advice that countries struggling with food security embrace massive one-crop farms as a viable alternative to small, biodiverse systems. This clip provides opportunity to discuss the issues that arise when humans allow corporate interests to manage local food quality and distribution.

The Big Ag Conference—47:08

In this scene, Evans and Reade attend a conference on edible insects where researchers and chefs are lobbied by food corporations already eyeing insects as a marketable protein source to invest in short-term profit rather than long-term food availability. Reade, feeling defeated after the conference, laments, "It's about making the system they already have more sustainable rather than recognizing the whole . . . system they've based their economy on is just going to destroy the place."[24] At this point, students know and like Evans and Reade and empathize with their frustration. Some will question how incorporating insects into the Western diet could make current agricultural practices more sustainable. This is a good time to discuss how the film models standard research practice, which often involves predicting one outcome and being surprised by another. Students could then explore how the researchers' honesty, humility, and vulnerability help build their ethos and the film's intent.

Evans's Final Assessment—1:09:15

In the film's final scene, Evans argues that the FAO's findings may be flawed since another arm of the UN, the World Food Program, recognizes that there is currently enough food to feed twelve billion people. Evans suggests that more attention should be paid to "the power structures that perpetuate inequality of access to food"[25] rather than creating a market niche for insect-based cuisine. The final scene does not offer definitive solutions to the issues presented in the documentary, perhaps implying that viewers will be wrestling with them for years to come. Here, students can be led to brainstorm how developing environmentally friendly food chains on the local level may lessen the effect of regional and global food crises.

After large-group discussion, students should be ready to answer the key questions that drive a film review: what does the film attempt to do, and how well does it succeed? Some students may claim the film tries to convince Western viewers to consume insects—a good place to pause and reinforce the broader climate change issues explored during prior discussion. Students should be led to understand the film's goal is to allow the viewer to consider how they may unknowingly contribute to climate change problems through overconsumption and suggests critical and conscious actions that can lead to a sustainable food system. While some students claim they are unwilling to try insects, most express gratitude for having watched the film as it provides a new perspective on the daily dietary habits they take for granted.

Three first year writing students' reactions are as follows:

Student one: "I was like most Americans [disgusted] at the idea of eating bugs but . . . [after viewing the film] . . . I was honestly wanting to try them. I would recommend the film *Bugs* [to] everyone . . . [because the filmmakers] . . . are trying to get people to understand the problem we will be facing sooner than later. [Eating insects] would most certainly be healthier and would provide [viewers] with more food options that are simply found in nature."

Student two: "The film is certainly successful in showing . . . the real, raw reactions of the team as they capture and cook bugs. I believe that this use of honest reactions was necessary for the film to be as successful as it was in getting its message across, the message that we don't have all the answers. The truth . . . is that no one really knows what the answer is for the problems of the future. All we can do now, and what they have done, is find possible solutions. . . . "

Student three: "Students should watch this film because we are adults emerging into the 'big world' . . . right now, we don't have the complete authority . . . on this issue; however, it will be up to students to deal with [climate change] and find something that will save the planet. Though I agree that some scenes do provoke a disgusted response [in viewers] . . . the big idea here is to search for other alternative food sources with a creative mind."

These reflections clearly indicate students are interested in and open to exploring solutions to the issues explored in the film. Yet, despite having a clear sense of urgency regarding climate change, many students feel they lack the expertise and training to address such concerns. After one viewing during the spring semester, a full year into the pandemic, one student remarked during class discussion that for years her teachers had apprised students of the growing number of climate change threats, but that viable solutions for these concerns were rarely if ever addressed. The problem became so large in the student's mind that, while she earnestly believed the reality of climate change should be clarified and confronted, she doubted any organization or government entity would or even could begin to mitigate global temperature rise's deleterious effects. She recognized that many educators and experts spoke publicly about climate change and deftly critiqued the unwillingness of powerful stakeholders to act. But from her perspective, influential adults willing to talk about climate issues were also absolving themselves of immediate responsibility while passing the buck to future generations.

Given the above-mentioned feedback, we feel an even greater need to educate students on how to take charge of their own learning. As general education instructors, we are aptly positioned to show them how and where they can find accessible, reader-friendly research on climate related issues that they can read, evaluate, and discuss with the goal of implementing a climate-change action plan tailored to their personal needs and expectations. Forister, Pelton, and Black's "Declines in Insect Abundance and Diversity: We Know Enough to Act Now" is a source that recommends specific solutions to insect population decline that are doable by nonexpert and nonscientific audiences. The authors clearly remind that "society must take steps at all levels to protect, restore, and enhance habitat for these animals across all landscapes, from wildlands to farmlands to urban cores" and lists actionable steps governments, landowners, and land managers can take to help protect existing habitats, move away from unsustainable agricultural practices, and encourage everyday citizens to promote insect biodiversity.[26] While students who want to take part in the changes implemented to reduce the negative impact of climate change may feel they lack the resources or ability to do so, this article recognizes that everyone, whether homeowners or renters, can make a difference by planting inground or container mini-gardens, reducing or ending the use of pesticides, buying foods grown with organic or sustainable methods, and advocating that governments protect and maintain natural land and water spaces noting "even a backyard or apartment balcony can be an important stopover for the smallest of animals upon which we all depend."[27] Resources like this should be shared with students and nonexperts, providing the permissions they need and want to mitigate the negative impacts of climate change whenever and wherever possible. The scientific community has long been aware that "acting with imperfect knowledge is something we do all the time, in our personal and professional lives."[28] Knowing this gives students the agency to determine for themselves how to make environmental diversity and sustainability a long-term goal.

As previously noted, Cherry's article offers important suggestions on and motivation for reducing biases on complex climate change issues that may leave teachers, as well as students, feeling threatened or overwhelmed. First, Cherry reminds that simply having an awareness of bias helps people understand how it influences their thinking.[29] Second, the author suggests that readers consider what "factors may influence [their] thinking by posing reflective questions like 'Are there factors such as overconfidence or self-interest at play?'" when one is making a decision.[30] Finally, Cherry calls on readers to challenge their biases once they recognize them, arguing that "thinking about [shortcuts in one's thinking] and challenging . . . bias can make [one] a more critical thinker."[31]

BEHIND THE CURVE IN THE PHYSICS CLASSROOM

Behind the Curve centers on the main actors of the flat-Earther community as they conduct experiments designed to (unsuccessfully) prove the Earth is flat and create a visible platform for their agenda. While the film does not directly address climate change, it creates a solid parallel to the origins of climate change deniers' giving explanation to the development and confirmation of their misguided beliefs, while suggesting that the exclusive discourse of the scientific community may bear some responsibility in the public's growing rejection of science as the guardian of truth. The film advocates that scientists have a responsibility to engage in more community outreach in order to develop a society that values both science and all its members. It is our belief that the larger problem is a matter of a decline in discourse and our society's ability to communicate and listen in a constructive and productive manner. We have all witnessed an extreme ideological polarization in the public sphere. Science and academia have become associated with political ideology rather than lenses by which we pursue objectivity together through our limited subjective experiences. Misinformation is rampant throughout social media and has become so widespread that mistrust in expertise is common-place and exacerbated by everyone having access to the publication of their opinion. Addressing these social issues in the classroom is crucial to affecting social change and documentaries on climate change provide a narrative that appeals to the human condition. The act of listening and watching as a story unfolds engages the classroom in social discourse and invites them to share in a community of practice guided by expertise. While documentaries provide this opportunity in any classroom, it is less commonplace that the opportunity for such discussions occur in science classrooms. Science classrooms most commonly implement a traditional lecture format. In 1998 Hake utilized pre/post Force Concept Inventory scores from 6,000 students across sixty-two physics courses, comprising high school, college, and university populations, and found learning gains in courses where "interactive engagement methods" were implemented to be consistently higher than courses defined as "tradi-tional lecture."[32] Documentary films as text provides an accessible implemen-tation of interactive engagement for science instructors seeking to transform their classroom to a more student-centered pedagogical space.

Physics students are given a brief introduction to the film *Behind the Curve* along with a rationale for viewing it. It is important to preface that even though the film explores controversial beliefs, the intent of watching it as a class is not to mock individuals who subscribe to conspiracy theories. Instead, we might consider the ability to understand and engage with others who hold different beliefs as an important skill for scientists and stipulate that it is human nature to have cognitive bias. In this case, both film stud-ies discussed have been implemented on college campuses in Missouri that serve populations from smaller rural communities. Valdez et al. determined students were more likely to engage in behaviors to adapt to or mitigate cli-mate change when they were from urban, high-socioeconomic-status schools and less likely to engage in these behaviors if they were from urban, low-socioeconomic-status schools or rural schools.[33] In *Don't Even Think About It*, George Marshal provides evidence that attitudes toward climate change can be predicted by cultural characteristics and sociopolitical demographics.[34] It is important to recognize the possibility that a student population dominated by first-generation students in rural Missouri and similar locations may contain members that identify, relate to, or recognize their own family and friends in the characters heavily invested in conspiratorial theories. Marshall asserts a clear lesson from the cultural clash created by climate change sci-ence: "communications from people's family, friends, and those they regard as being like themselves (their peers) can have far more influence on their views than the warning of experts."[35] Therefore, any discussion surrounding characters engaged in climate denial will be more productive if the instructor explicitly shows these characters empathy and facilitates an informed discussion on climate science among peers in the classroom.

Showing these characters' empathy shows empathy by proxy to students more likely to feel they are not welcome in an academic discussion on climate change, while modeling constructive discourse behaviors. I share my own experience with friends and family members who believe varying conspiracy theories such as "The moon landing was fake," or "Elvis never died," or "Climate change is a hoax." I explain how I learned that listening to these positions without judgment and exploring ideas, even controversial ones, was always more productive than immediately discrediting or mocking them. As a result, I was able to engage in continuing discussion that built trust, which allowed me to share resources and information. Finally, before starting the first clip, I present the question, "How are you a flat-Earther?," to remind students that even those with whom we strongly disagree deserve our empathy.

This documentary is not used to introduce the concepts relative to understanding climate change, but to recognize the cognitive shift that must take place between understanding and denialism for flat-Earthers to argue against scientific theory and the models used to accurately predict phenomena in the universe. Conspiracy theories have common features that are necessary to sustain them in the face of overwhelming evidence to the contrary. The documentary explicitly addresses the difference between skepticism and denialism. Skepticism is essential to the scientific process and scientific community as it encourages and informs peer review. Denialism assumes that neither evidence nor entity can alter the denier's core beliefs and foundational principles.

Learning is commonly viewed as a negotiation between student's current ideas or beliefs and what is being presented in the classroom. Students are not empty vessels into which knowledge is poured. Posner et al. views learning as a kind of inquiry and like inquiry "learning is a process of conceptual change."[36] These purely cognitive models of learning have been further informed by the sociocultural lens of participationism. Rogoff describes learning as participation.[37] Rogoff posits that due to constantly participating in social interactions, a person's participation continually feeds back into their understanding and results in a transformation of how they participate.[38] Learning is then exhibited through new forms of participation within communities. Denialism and skepticism are two distinct ways to participate in the conversation on climate change. Creating parallels between flat-Earthers and climate change deniers provides opportunity to effect conceptual change and transform how these students participate in the discussion of climate change.

Posner et al. established four steps in affecting conceptual change.[39] The first step requires that students become dissatisfied with their existing conceptions as explanations for data.[40] Before presenting the scientific models that explain climate change, developing clear parallels between flat-Earthers and climate change deniers creates cognitive dissonance in students whose thinking aligns with denialism. Denying the Earth is round requires that flat-Earthers distrust NASA's extensively documented orbital research and vilify scientists who analyze and support NASA's findings. This also holds true for climate change deniers who question the validity of data that overwhelmingly supports the correlation between a rise in atmospheric carbon dioxide levels and an increase in average global temperatures. The second step requires the new conception to be "intelligible."[41] The documentary provides multiple examples of the flat-Earth community conducting well designed experiments to test their hypothesis. In each of these experiments the data contradicts their hypothesis, but the flat-Earth community consistently denies the data as evidence of a round Earth. Students watch in real time as flat-Earthers perform mental gymnastics in desperation to assert they are still correct in the face of proving themselves clearly wrong. The third step requires the new conception to be "initially plausible."[42] As more and more data are available to the public, including recent civilian missions to space, it is broadly accepted by the student population that the Earth is in fact round. The fourth step requires the new conception suggest the "possibility of a fruitful research program."[43] The documentary establishes a clear model for what denialism is in practice. The arguments made by the members of the flat-Earth community are addressed by the scientists in the film providing students with academic language and theory to justify their disagreement with the flat-Earth theory. After developing an understanding of denialism in the context of flat-Earthers, transitioning to a discussion on the science of climate change helps students confront the irrefutable data. There is a precedent established that skepticism is encouraged, but our ideas are malleable to hard evidence.

What follows is the recommended sequence of clips along with discussion questions for students.

Introduction to the Flat-Earther Theory—3:40–6:04

In this scene, viewers are introduced to a prominent flat-Earther, Mark Sargent, who illustrates the flat-Earth assumption that he should not be able to view the Seattle skyline at a distance. He then correctly models the curvature of the Earth in a representation that explains why objects at a distance are not in full view and then predicts that he should not be able to see Seattle's most prominent building from his side of the lake. Because of the Earth's curvature, however, Sargent not only sees the tallest building across the lake, he can see the upper halves of other buildings.

Students should recognize Sargent's native intelligence and curiosity. He models the curvature of the Earth to test its validity but is unwilling to accept data from his experiment. When asked what takeaways they have after viewing the clip, many students discuss Sargent himself and his relationship with his mother, who is also introduced in the clip since they live together, expressing empathy for his willingness to share his homelife. Other students, shocked that Sargent believes the Earth is flat and that there is a flat-Earther community, begin to interrogate his logic.

What Is Science?—25:51–29:32

In this clip, scientists explain two types of cognitive bias, imposter syndrome and the Dunning–Krueger effect, both of which offer a scientific model of thinking that leads to anti-intellectualism and conspiracy theories, such as that espoused by flat-Earthers. A scientist provides a definition of science, and Sargent verifies that no scientist has yet come forward as a flat-Earther. After viewing this clip, students should be reminded of the definition of a theory: we cannot prove anything true; we can only prove something to be false. One definition of scientific theory is a well-constructed and peer-reviewed set of experiments and data that consistently cannot be proven false when replicated. Allow students to ask questions and suggest that this clip helps the audience understand how cognitive bias occurs and reaffirms the role of expertise.

Falsifiable Hypothesis—46:51–51:01

The filmmakers pose a question: is there any evidence that would convince you to change your mind on the flat-Earth theory? In the documentary, a scientist explains that one's understanding is not scientific if a hypothesis is not verifiable. Next, several flat-Earthers design a brilliant experiment using a laser gyroscope that is capable of detecting the fifteen-degree-per-hour drift caused by a spherical Earth's 360-degree rotation during a twenty-four-hour period. They hypothesize that if the gyroscope does not measure a fifteen-degree drift, then the Earth must be flat. However, the instrument does measure a fifteen-degree drift, which ultimately supports the view that the Earth is spherical. Despite evidentiary proof, the flat-Earth spokesman still does not accept the results and instead invokes more complicated parameters to support his assumptions. Students can discuss that there is no evidence that would change the flat-Earther's mind, which is consistent with the documentary's definition of denialism. Skepticism, on the other hand is crucial to the scientific process and the documentary argues that flat-Earthers have this quality or the instincts of a good scientist. The problem lies in their methodology: they start with a conclusion and cherry-pick evidence to support it. Next, students discuss the experimental method and data analysis. Science is a process that does not begin with a result; rather, a question is posed and then data is collected to develop an understanding of the physical phenomenon.

Dismantling the Scientific Superiority Complex—1:08:25–1:13:48

Several flat-Earthers open up about losing friends, family, and their support systems because of their beliefs. Then, the film cuts to a scientific conference where one scientist addresses the rise of conspiracy theories in mainstream culture. He argues that it is the responsibility of the scientific community to prevent the alienation of would-be scientists who are seeking community. Students explore the responsibilities of those in authority positions, including how empathy and humility are important traits to develop. Students also explore the notion that all humans are capable of cognitive bias. Students are prompted to ask: how am I a flat-Earther? Many admit they have friends and family who believe in conspiracy theories while others claim they have a hard time relating to flat-Earthers.

One Last Experiment—1:33:00–1:35:30

A flat-Earther revises his experiment from earlier in the documentary, but the results are the same: his evidence supports a spherical Earth. Still, he finds a way to create doubt and refuses to accept the results, believing instead that an unforeseen error occurred to alter the outcome he anticipated.

Immediately after the film's final scene, offer students an opportunity to make connections between the film and climate change: "Now let's consider climate change. We are going to discuss the science that allows us to understand how the data show a direct correlation between rising global average temperatures and an increase in CO_2 concentration in the atmosphere. We will look at the main contributors to the growing CO_2 concentration and the effects we experience to rising global temperatures." Students are engaged and entertained by the outlandish characters and ideas presented in the film. Their engagement is evident as students excitedly react and share their confusion about how people could disagree with something like the shape of the Earth that they see as so elementary. I encourage their ideas and participation but remind them of the movie's stance on its subjects: flat-Earthers are not unintelligent people but misinformed and often disenfranchised. I reinforce the role of humility in expertise. I remind them that science is a community of practice that is responsible for being skeptical and seeking consensus. This creates space for a discussion about climate change that focuses on understanding scientific principles and respecting the rigorous process by which theory is tested and applied.

As climate data are presented, parallels are established between flat-Earther and climate- denialist beliefs that require NASA and scientists to be lying, conspiratorial, and corrupt. Opportunity should be allowed for students to ask questions about how data are taken, and it is important to discuss why data are reliable. This strategy has led to fantastic discussions with students who are skeptical of climate change and promoted transformative engagement. One anecdotal example: a student vehemently opposed the notion that climate change was real. He quoted common arguments and provided "sources," such as blog posts and papers by active climate deniers. His opposition prompted one-on-one discussions that pointed out errors, blatant fabrication, and sources that were not published in scientific journals that allow for academic scrutiny. By semester's end, the student openly expressed a refined opinion and conceded that climate was changing, though remained unconvinced that it was due to human industry. I encourage students to watch the full documentary on their own and am gratified to hear them report watching and discussing it with family, friends, and workmates.

CONCLUSIONS

We believe the films' compelling and confounding subject matter—edible insects and conspiracy theories—provide a hook for students to enter into real conversations about climate-change issues. Students become excited sharing their reactions to the films and are willing to explore deeper issues with their peers. We believe using film as text encourages in-the-moment critique during which students share first impressions, pose questions, and engage in friendly, low-stakes debate. As general education instructors we reach a wide range of students with different interests, backgrounds, and professional goals. We feel compelled to integrate climate change discussion into our curriculum and find documentary film to be an efficient and persuasive text to share with students as its impact seems to extend beyond the walls of our classrooms.

NOTES

1. Shu-Chiu Liu, "Environmental Education through Documentaries: Assessing Learning Outcomes of a General Environmental Studies Course," *Eurasia Journal of Mathematics, Science and Technology Education 14*, no. 4 (2018): 1371.

2. Bernard R. Robin, "Digital Storytelling: A Powerful Technology Tool for the 21st Century Classroom," *Theory into Practice* 47, no. 3 (2008): 220–28.

3. Veleda Boyd and Marilyn Robitaille, "Composition and Popular Culture: From Mindless Consumers to Critical Writers," *The English Journal* 76, no. 1 (1987): 51.

4. Jessie Daniels, "Transforming Student Engagement through Documentary and Critical Media Literacy," *Theory in Action* 5, no. 2 (2012): 4–29.

5. Alan S. Marcus and Jeremy Stoddard, "The Inconvenient Truth about Teaching History with Documentary Film: Strategies for Presenting Multiple Perspectives and Teaching Controversial Issues." *The Social Studies* 100, no. 6 (2009): 279–84.

6. Liu, 1371.

7. John M. Dirk, "The Power of Feelings: Emotional, Imagination, and the Construction of Meaning in Adult Learning." *New Directions for Adult and Continuing Education* 89 (Spring 2001): 63.

8. Saffron O'Neill and Sophie Nicholson-Cole. "'Fear Won't Do It': Promoting Positive Engagement with Climate Change through Visual and Iconic Representations," *Science Communication* 30, no. 3 (2009): 373.

9. Rene X. Valdez, M. Nils Peterson, and Kathryn T. Stevenson, "How Communication with Teachers, Family and Friends Contributes to Predicting Climate Change Behaviour among Adolescents." *Environmental Conservation* 45, no. 2 (2018): 190.

10. J. A. Thomas, N. A. D. Bourn, R. T. Clarke, K. E. Stewart, D. J. Simcox, G. S. Pearman, R. Curtis et al., "The Quality and Isolation of Habitat Patches Both Determine Where Butterflies Persist in Fragmented Landscapes," *Proceedings. Biological Sciences* 268, no. 1478 (2001): 1791–96.

11. Caspar A. Hallmann, Martin Sorg, Eelke Jongejans, Henk Siepel, Nick Hofland, Heinz Schwan, Werner Stenmans et al. "More Than 75 Percent Decline over 27 Years in Total Flying Insect Biomass in Protected Areas." *PLoS ONE* 12, no. 10 (2017): 1–21.

12. M. L. Forister, E. M. Pelton, and S. H. Black, "Declines in Insect Abundance and Diversity: We Know Enough to Act Now," *Conversation Science and Practice* 1, no. 80 (2019): 6.

13. Forister, Pelton, and Black, 1.

14. J. C. Biesmeijer, S. P. M. Roberts, M. Reemer, R. Ohlemuller, M. Edwards, T. Peeters, A. P. Schaffers, A.P., "Parallel Declines in Pollinators and Insect-Pollinated Plants in Britain and the Netherlands," *Science*, July 21, 2006: 351.

15. Kendra Cherry, "What Is Cognitive Bias?" *Verywell Mind*, July 19, 2020, accessed December 29, 2021, http://www.verywellmind.com.

16. Cherry.

17. Cherry.

18. Cherry.

19. Dan Ariely, *Are We Really in Control of Our Decisions?* TED, 2009. YouTube.

20. Cherry.

21. Cherry.

22. Andreas Johnsen, dir. *Bugs.* Danish Film Institute, 2016. YouTube.

23. Bugs.

24. Bugs.

25. Bugs.

26. Forister, Pelton, and Black, 3–4.

27. Forister, Pelton, and Black, 6.

28. Forister, Pelton, and Black, 6.

29. Cherry.

30. Cherry.

31. Cherry.

32. Richard R. Hake, "Interactive-Engagement Versus Traditional Methods: A Six-Thousand-Student Survey of Mechanics Test Data for Introductory Physics Courses." *American Journal of Physics* 66, no.1 (1998): 2.

33. George Marshall, *Don't Even Think About It: Why Our Brains Are Wired to Ignore Climate Change*. (New York: Bloomsbury Publishing, 2015).

34. Marshall, 24.

35. G. J. Posner, K. A. Strike, P. W. Hewson, and W. A. Gertzog, "Accommodation of a Scientific Conception: Towards a Theory of Conceptual Change," *Science Education 67* (1982): 212.

36. B. Rogoff, "Explanations of Cognitive Development through Social Interactions: Vygotsky and Piaget," in *Apprenticeship in Thinking: Cognitive Development in Social Context.* (Oxford University Press, 1990): 1–16.

37. Rogoff, 1–16.

38. Posner et al., 214.

39. Posner et al., 214.

40. Posner et al., 214.

41. Posner et al., 214.

42. Posner et al., 214.

43. Posner et al., 214.

BIBLIOGRAPHY

Ariely, Dan. *Are We Really in Control of Our Decisions?* TED, 2009. YouTube.

Biesmeijer, J.C., S. P. M. Roberts, M. Reemer, R. Ohlemuller, M. Edwards, T. Peeters, A. P. Schafferset al. "Parallel Declines in Pollinators and Insect-Pollinated Plants in Britain and the Netherlands." *Science* (July 21, 2006): 351–54.

Boyd, Veleda, and Marilyn Robitaille. "Composition and Popular Culture: From Mindless Consumers to Critical Writers." *The English Journal 76*, no. 1 (1987): 51–53.

Cherry, Kendra. "What Is Cognitive Bias?" *Verywell Mind*, July 19, 2020. Accessed December 29, 2021. http://www.verywellmind.com.

Clark, Daniel J., dir. *Flat Earthers: Behind the Curve.* Netflix, 2018.

Daniels, Jessie. "Transforming Student Engagement through Documentary and Critical Media Literacy." *Theory in Action 5*, no. 2 (2012): 4–29.

Dirk, John M. "The Power of Feelings: Emotional, Imagination, and the Construction of Meaning in Adult Learning." *New Directions for Adult and Continuing Education* 89 (spring 2001): 63.

Dole, Janice A., and Gale M. Sinatra. "Reconceptualizing Change in the Cognitive Construction of Knowledge." *Educational Psychologist* 33, nos. 2–3 (1998): 109–28.

Forister, M. L., E. M. Peltone, and S. H. Black. "Declines in Insect Abundance and Diversity: We Know Enough to Act Now. *Conservation Science and Practice* 1, no. 80 (2019): 1–8. Accessed December 29, 2021. https://doi.org/10.1111/ csp2.80.

Hake, Richard R. "Interactive-Engagement versus Traditional Methods: A Six-Thousand-Student Survey of Mechanics Test Data for Introductory Physics Courses." *American Journal of Physics* 66, no.1 (1998).

Hallmann, Caspar A., Martin Sorg, Eelke Jongejans, Henk Siepel, Nick Hofland, Heinz Schwan, Werner Stenmans et al. "More Than 75 Percent Decline over 27 Years in Total Flying Insect Biomass in Protected Areas." *PLoS ONE* 12, no. 10 (2017). Accessed December 29, 2022. https://doi.org/10.1371/journal. pone.0185809.

Johnsen, Andreas, dir. *Bugs*. Danish Film Institute, 2016. YouTube.

Liu, Shu-Chiu. "Environmental Education through Documentaries: Assessing Learning Outcomes of a General Environmental Studies Course." *Eurasia Journal of Mathematics, Science and Technology Education* 14, no. 4 (2018): 1371–81.

Marcus, Alan S., and Stoddard, Jeremy. "The Inconvenient Truth about Teaching History with Documentary Film: Strategies for Presenting Multiple Perspectives and Teaching Controversial Issues." *The Social Studies* 100, no. 6 (2009): 279–84.

Marshall, George. *Don't Even Think about It: Why Our Brains Are Wired to Ignore Climate Change*. New York: Bloomsbury Publishing, 2015.

O'Neill, Saffron, and Sophie Nicholson-Cole. "'Fear Won't Do It': Promoting Positive Engagement with Climate Change through Visual and Iconic Representations." *Science Communication* 30, no. 3 (2009): 355–79.

Posner, G. J., K. A. Strike, P. W. Hewson, and W. A. Gertzog. "Accommodation of a Scientific Conception: Towards a Theory of Conceptual Change." *Science Education* 67 (1982): 211–27.

Robin, Bernard R. "Digital Storytelling: A Powerful Technology Tool for the 21st Century Classroom." *Theory into Practice* 47, no. 3 (2008): 220–28.

Rogoff, B. "Explanations of Cognitive Development through Social Interactions: Vygotsky and Piaget," in *Apprenticeship in Thinking: Cognitive Development in Social Context*. Oxford University Press, 1990: 1–16.

Thomas, J. A., N. A. D. Bourn, R. T. Clarke, K. E. Stewart, D. J. Simcox, G. S. Pearman, R. Curtis, and B. Goodger. "The Quality and Isolation of Habitat Patches Both Determine Where Butterflies Persist in Fragmented Landscapes." *Proceedings. Biological Sciences* 268, no. 1478 (2001): 1791–96.

Valdez, Rene X., M. Nils Peterson, and Kathryn T. Stevenson. "How Communication with Teachers, Family and Friends Contributes to Predicting Climate Change Behaviour among Adolescents." *Environmental Conservation* 45, no. 2 (2018): 183–91.

Chapter Three

Educating Space-Age Environmentalists at the Elementary Level

Beverly B. Bachelder and Robert S. Bachelder

The director of the United Nations Office for Outer Space Affairs, Simonetta di Pippo, recently called for immediate action by the international community to address the dual threats of climate change and space debris. Scientists' warnings about the dangers posed by these environmental problems have been ignored for too long, the Italian astrophysicist said, and it is time to listen to science before it is too late.[1]

It also is time to educate a new generation of environmentalists who are equipped to address the dual threats of climate change and space debris. Our planet needs space-age environmentalists whose perspective encompasses Earth and space, who are both climate literate and space literate, who understand the connection between a sustainable planet and a sustainable space environment, and who are motivated to take age-appropriate action. When we speak of "space" in this context, we refer specifically to that region of space known as near-Earth orbit where satellites orbit the planet to provide a wide array of benefits to people on Earth. It is "a global common and a limited resource," Director di Pippo says, and "we must protect it for future generations."[2]

This chapter fills a serious gap in environmental education by introducing a new topic that we call "space literacy." Climate literacy is a familiar topic. Students become climate literate as they learn how climate affects society, how human activity impacts climate, and how society can make informed and responsible decisions regarding climate. We frame the topic of space literacy in a complementary way. Students become space literate as they learn how space affects society, how human activity impacts space, and how society can make informed and responsible decisions regarding space.

Three essential principles serve as a framework for instruction in space literacy: (1) The orbital space environment is a limited resource that is home to spacecraft we depend on for daily life and solving global problems, especially climate change; (2) human activity in space creates an orbiting junkyard of debris that threatens spacecraft and astronauts; and (3) solving the debris problem requires new technologies and citizen advocacy for new rules and international agreements. We consider each of these principles in turn.

Orbital space is a limited natural resource that serves as home to the International Space Station, the Hubble space telescope, and over 5,800 active satellites owned by governments, commercial companies, and non-profits.[3] Space-based technologies are so ubiquitous in our daily lives that we tend to take them for granted. They are essential for phone communications, credit card authorization networks, video feeds for cable and broadcast transmissions, weather forecasts, and GPS (global positioning system) navigation.

Satellites also provide indispensable support for global efforts to address climate change. NASA's fleet of Earth observation satellites, for example, is designed for long-term global observation of the climate system including the atmosphere, biosphere, cryosphere, land surface, and oceans. These and other satellites provide scientists and policymakers with authoritative and essential information for modeling the evolution of climate, devising mitigation and adaptation strategies, and assessing risks related to famines, water shortages, and natural disasters. Geographer Waleed Abdalati says, "a loss of our observational capabilities would be like closing our eyes, handicapping our ability to know what tomorrow, next week or next decade will bring."[4]

The Aqua satellite's Clouds and the Earth's Radiant Energy System (CERES), for example, measures reflected sunlight and thermal radiation emitted by Earth so that scientists can understand how our planet's heat budget is changing over time. The Aura satellite's instruments, including its Ozone Monitoring Instrument (OMI) and Microwave Limb Sounder (MLS), measure ozone, aerosols, and gases to help scientists analyze Earth's atmosphere, including the impact of the Amazon rainforest on its chemical composition. The Sentinel-6 Michael Freilich satellite uses its radar altimeter to provide researchers with invaluable information about global and regional changes in sea level and their implications for a warming world. The Moderate Resolution Imaging Spectroradiometer (MODIS) on board the Terra satellite, the flagship of NASA's Earth observation fleet, provides a global perspective on the relationship between climate change and agricultural production for policymakers trying to bolster agricultural productivity and plan for potential famines.

The International Space Station (ISS) also provides critical information for understanding and addressing environmental issues. Passing over 95 percent of the Earth's inhabited surface every twenty-four hours, the ISS provides a unique vantage point for observing Earth in real time with hands-on equipment. Astronauts collect camera images that facilitate rapid response to floods, fires, earthquakes, volcanic eruptions, deforestation, harmful algal blooms and other types of natural events.

To appreciate why space cannot weather mistreatment any better than the atmosphere or oceans, it is important to understand how the resource system functions in space. Satellite operators use common engineering solutions based on orbital mechanics that lead to a clustering of spacecraft in orbits that fit their missions. There are several types of Earth orbits and each provides operators with distinctive capabilities.[5]

Most active satellites reside in low Earth orbit (LEO), the region of space extending 1,200 miles (2,000 kilometers/km) above Earth's surface where it takes them between ninety minutes and two hours to complete a full orbit. The combination of close proximity to Earth and short orbital periods makes LEO ideal for Earth observation missions.

Satellites in geostationary orbit (GEO) operate at an altitude of 22,300 miles (35,900 km) above the equator. They travel from west to east at the same speed as the Earth's rotation and appear to be in a fixed position. GEO permits communications satellites to transmit a signal to an antenna in a fixed position on the ground and also is used by weather satellites that continually monitor trends in a specific geographic area.

Medium Earth orbit (MEO), the region between LEO and GEO, is home to GPS satellites that orbit at an altitude of approximately 12,550 miles (20,200 km). They send their signals to a receiver in a telephone or vehicle. Once the receiver calculates its distance from four or more satellites, it can pinpoint the user's location.

Because of this clustering of spacecraft, experts say the resource system in space resembles a parking garage where the basic resource units are parking spaces and the number of spaces is limited.[6] The "parking garage effect," as we call it, is a foundational concept for space literacy in much the same way the greenhouse effect is a foundational concept for climate literacy. The greenhouse effect explains how carbon dioxide and other greenhouse gases keep Earth warm enough to sustain life by trapping heat from the sun before it escapes back into space. The parking garage effect explains how orbital space functions to provide a wide range of benefits to society, from Earth observation to communications and navigation.

Physical crowding issues with cars and SUVs make it a challenge to manage a parking garage. It is a similar situation in orbital space where the number of slots is limited. In space, though, the situation is more complicated and dangerous. Imagine what it would be like to share your local parking garage not only with other drivers but with vehicles that had broken down in the garage years ago and were never removed. This is the situation in space where active spacecraft must crowd into orbits with broken down vehicles and junk parts from old missions.

The expanding utilization of space has created major challenges for the international community in its stewardship of the space environment. Key orbits, especially in LEO below 620 miles (1000 km) where Earth observation satellites reside, have become congested with debris.[7]

Routine space operations create debris. Large debris objects include spent rockets and over 2,600 defunct satellites.[8] Smaller objects include stray bolts and screws, astronauts' tools, gloves, discarded camera lens caps, and paint flecks. Additional debris has been generated by over 600 explosions, collisions, anti-satellite weapons tests, and other events.[9]

Space debris travels at speeds of up to 18,000 miles per hour (28,968 km), almost seven times faster than a bullet.[10] NASA reports there are millions of pieces of orbital debris in LEO with at least 26,000 of them the size of a softball or larger that could "destroy a satellite on impact; over 500,000 the size of a marble big enough to cause damage to spacecraft or satellites; and over 100 million the size of a grain of salt that could puncture a spacesuit."[11]

NASA characterizes the LEO environment in sobering terms:

> Spent rockets, satellites and other space trash have accumulated in orbit increasing the likelihood of collision with other debris. Unfortunately, collisions create more debris creating a runaway chain reaction of collisions and more debris known as the Kessler Syndrome after the man who first proposed the issue, Donald Kessler. It is also known as collisional cascading. . . . Kessler demonstrated that once the amount of debris in a particular orbit reaches critical mass, collision cascading begins even if no more objects are launched into the orbit. Once collisional cascading begins, the risk to satellites and spacecraft increases until the orbit is no longer usable.[12]

According to studies by NASA and other space agencies, debris has already reached critical mass in heavily used orbits in LEO.[13] Moreover, experts estimate that some 50,000 satellites could be launched into orbit during the next ten to fifteen years, with most of them crowding into LEO.[14]

Perhaps no one appreciates the threat posed by debris as acutely as astronauts on board the ISS. With increasing frequency, the ISS has been forced to fire its thrusters and perform collision avoidance maneuvers when debris passed too close for comfort. On several occasions, astronauts have been ordered by mission control to shelter in the Soyuz escape capsule docked outside the station in preparation for a quick getaway. Astronaut Scott Kelly, who sweated it out in the Soyuz during a close encounter with a fragment from a defunct Russian weather satellite, warns there is danger in becoming complacent about life in space.[15]

The "congested parking garage effect," as we call it, explains how the buildup of human-generated space debris creates an orbiting junkyard that threatens satellites and astronauts. It is the space counterpart to the enhanced greenhouse effect that explains how the buildup of human-generated greenhouse gases in the atmosphere causes global warming.

To address climate change effectively, we need to pursue a dual strategy of "reduce and remove." We need to reduce carbon emissions through carbon pricing, and we need to remove carbon dioxide from the atmosphere through reforestation and direct air capture technologies. We also need to pursue a dual strategy of "reduce and remove" for space debris. We need to reduce the volume of debris generated by new space missions, and we need to remove obsolete satellites and expended rockets from crowded orbits.

Guidelines to minimize the creation of new debris have been adopted by the United States and some other nations on a voluntary basis. For example, de-orbiting satellites from LEO at the end of their missions so they burn up in the atmosphere instead of adding to orbital congestion is of critical importance. Because this maneuver requires additional fuel and adds to the cost of missions, however, some countries and commercial firms are unwilling to follow the guideline.

Because debris is growing exponentially, removing dead satellites and rocket bodies from densely populated regions of LEO where collisions are most likely to occur is also essential for stabilizing LEO. This process would reduce the volume of existing debris from which new debris can be created and is called active debris removal, or ADR.

Space agencies and private companies around the world are researching and developing a variety of ADR concepts for in-orbit testing. Prospective technologies include lasers to slow down objects so that they fall into the atmosphere more quickly and burn up, electrodynamic tethers to generate drag and lower dead spacecraft into the atmosphere, space nets to capture debris, solar sails to accelerate the de-orbiting of satellites when they complete their missions, and "space garbage trucks" with robotic arms to rendezvous and grapple with obsolete satellites.

Raymond J. Sedwick, who directs the Space Power and Propulsion Laboratory at the University of Maryland, describes the challenges of grappling with inactive satellites: "Imagine trying to grab a spinning skater on an ice rink except instead of a person, it's an SUV traveling over 17,000 miles per hour. And instead of you being there in person, you're flying a drone, the lighting is bad, you have limited sensory data, there is no obvious place to grab onto, and you may be operating under a time delay."[16]

One impediment to capturing debris is that technologies that can grab objects on Earth do not work in space. Magnets, for example, do not stick to glass or aluminum which are commonly used materials in satellites. Conventional robotic hands cannot handle large, smooth pieces of space debris. One promising approach to overcoming this problem has been inspired by the setae and septulae on the bottom of geckos' feet that interact with surfaces through van der Waals forces. Scientists have developed gecko-mimicking adhesive structures that they place on the grippers used by robotic arms.[17] The grippers have been tested successfully in microgravity onboard the ISS with a variety of surfaces typically installed as part of satellites.

The political challenges associated with ADR also are thorny because the space garbage truck of one nation can be viewed as a potential weapon by an adversary. Unfortunately, diplomatic efforts to build the trust and confidence among major space powers that is needed for them to agree to remove old debris from orbit or to reduce the creation of new debris have not gotten off the ground. Solving such problems in space, observes James Clay Moltz of the Naval Postgraduate School, is similar to solving environmental problems on Earth, such as global warming. He writes: "We have learned a great deal about these shared risks, but self-interest and the growing fragmentation of power among countries, groups, and organizations make it difficult to reach a consensus and translate that into action."[18]

In 2021, NASA's Office of the Inspector General issued a report on the agency's efforts to address the orbital debris problem. It called on NASA and the United States to provide leadership that has been lacking to date. It recommended that NASA "lead national and international collaborative efforts to mitigate orbital debris including activities to encourage active debris removal and the timely end-of-mission disposal of spacecraft."[19] It further recommended that NASA "collaborate with Congress, other federal agencies, and partners from the private and public sectors to adopt national and international guidelines on active debris removal and strategies for increasing global compliance rates for timely removal of spacecraft at the end of a mission."[20]

US leadership on key policy issues would be welcome. Citizens should heed the advice of the Union of Concerned Scientists, a nonprofit organization that combines scientific research with public advocacy to address pressing global problems. Speaking of climate change—and the same can be said of space debris—it points out that "the best policy ideas in the world aren't worth much if we don't have activists, experts, and everyday people fighting for change. From school groups to churches; from corporate board rooms to mayors and local leaders: we need action."[21]

Elementary educators can take action by introducing the topic of space literacy with its three essential principles that provide a framework for instruction. Transmitting important information about the environmental crisis in space, however, is just the first step toward cultivating space age environmentalists. Rebecca L. Young, author and environmental education consultant, writes: "This preliminary goal of facilitating access to knowledge about environmental crises must lead to the action-oriented goal of nurturing students' commitment to empathy and initiative regarding them."[22] In a similar vein, an environmental literacy plan proposed for Massachusetts emphasizes the importance of providing students with "opportunities to experience natural systems in order to develop an emotional connection, appreciation, and motivation to protect them."[23]

Helping students recognize they are an integral part of nature and bear responsibility for its preservation and ongoing development presents an extra measure of difficulty when teaching about the space debris issue. As policy expert Thomas Krepon observes, "We are in the process of messing up space, and most people don't realize it because we can't see it the way we can see fish kills, algal blooms, or acid rain."[24]

Educators can address this impediment with a well-founded confidence in the power of books to connect students to the space domain. After all, it was a classic work of science fiction about a Martian invasion that inspired and motivated the scientific visionary who opened the door to the space age. On March 16, 1926, Robert H. Goddard successfully launched the world's first liquid-fueled rocket in Auburn, Massachusetts and watched it rise forty-one feet in 2.5 seconds before it hit the ground. According to NASA, the flight "was as significant to history as that of the Wright brothers at Kitty Hawk"[25] and Goddard's technology was used to propel astronauts to the moon in 1969. In 1932, Goddard wrote to author H. G. Wells:

> In 1898 I read your *War of the Worlds*. I was sixteen years old [and] it made a deep impression. The spell was complete a year afterward, and I decided that what might conservatively be called "high-altitude research" was the most fascinating problem in existence. The spell did not break, and I took up physics . . . how many more years I shall be able to work on the problem I do not know; I hope, as long as I live. There can be no thought of finishing, for "aiming at the stars," both literally and figuratively, is a problem to occupy generations.[26]

Literary texts that simultaneously instruct and inspire can provide much of the fuel that educators will need to achieve their mission objective. Jamil Zaki, a psychology professor and director of the Stanford Neuroscience Laboratory, notes that recent discoveries about regions of the brain that play an important role in empathy have changed his field's perspective on storytelling, a pastime that began when people first gathered around a fire and continues today by means of paper and screens. He writes:

Recently, psychologists have begun telling a new story about stories. More than a diversion, the narrative arts are an ancient technology: performance-enhancing drugs for untethering [the imagination]. Stories helped our ancestors imagine other lives, plan for possible futures, and agree on cultural values. In the modern world, they help in a new way: flattening our empathic landscapes, making distant others feel less distant and caring for them less difficult.[27]

Zaki points to research findings that young children who "devour story books" learn to identify with other people's emotions "sooner than their less bookish peers."[28]

Elementary level books with titles such as *Max Goes to the Space Station: A Science Adventure with Max the Dog* and *My Journey to the Stars* serve as literary lenses to help frame the essential principles of space literacy. They foster subject matter knowledge while connecting students to space at a personal level and helping them appreciate the importance of a clean space environment to their own lives and to the world. Stories told by videos on screens further deepen students' empathy for children who depend on satellite-based applications for basic necessities of life and for astronauts whose lives are threatened by space debris.

Together, stories on paper and screen equip and motivate students to undertake age-appropriate initiatives at the intersection of science and citizenship, effectively bridging the empathy-action gap. These initiatives incorporate English language arts, including drama, science, and fine arts, and tap students' imagination, ingenuity, and resourcefulness. They are vehicles that students can use to demonstrate their knowledge, educate their schools and communities about the environmental crisis unfolding in space, and take action now.

An award-winning book by Jeffrey Bennett titled *Max Goes to the Space Station* serves as a literary lens for teaching the first principle of space literacy: The orbital space environment is a limited resource that is home to spacecraft we depend on for daily life and solving global problems, particularly climate change. An astrophysicist and former NASA scientist and teacher, Bennett takes readers on a journey with Max the dog to the International Space Station. The book emphasizes the important role of the ISS in helping us study Earth from space.

Students enjoy listening to the book read aloud from the ISS by Astronaut Michael S. Hopkins via the Story Time from Space website.[29] Hearing the story in this way connects students to orbital space with a sense of immediacy as they are encouraged to care responsibly for our planet. The book is a prequel to other volumes in the Max Science Adventure Series, all of which represent a rare and effective blend of fiction and science. Students enjoy the beautifully illustrated, inspirational story while a series of "Big Kid Box" sidebars explains the science behind the story.

Max's adventure begins when he is selected as the first dog to visit the ISS as a space tourist. Prior to launch, he participates in a training program at the Johnson Space Center in Houston. Once in space, he experiences weightlessness and learns firsthand about life on board the ISS. As he watches astronauts conduct scientific experiments in the microgravity environment, the crew is suddenly faced with a life-threatening situation. The cooling system begins to leak and releases poisonous ammonia gas. Luckily, Max's insistent barking alerts the crew to the danger. Mission Commander Grant repairs the leak and saves the crew, and Max is celebrated as a hero.

Later, Commander Grant takes Max to the Cupola, the station's "room with a view." Bennett writes: "The view was spectacular, and it seemed to change everyone who ever saw it. After all, this was the *International* Space Station, where people from all over the world worked together in peace. Was it really so difficult to imagine that people could work together the same way on the beautiful blue world below?"[30]

The view from space is inspirational and also invaluable for generating much of the information and scientific perspective we need to work together to preserve our beautiful blue world. A sidebar, "Studying Earth from Space," features an image of Hurricane Irene and describes how the ISS and other satellites use their global overview of Earth to do science that cannot be done from the ground.[31] Students learn that spacecraft use special instruments that look down at Earth and measure temperatures, wind speeds, ocean currents, and other factors. This information helps scientists assess the impact of global warming caused by human activities and better understand how our planet works.

When students are asked what they know about space, they often mention Mars rovers, space shuttles, and robotic probes that explore the solar system. *Max Goes to the Space Station* helps students envision their relationship to space in a new light as they learn how life "up there" on the ISS benefits life "down here" on Earth. The central lesson of the story is encapsulated in its concluding paragraph: "Working in space may someday help us learn to live on other worlds, but its greatest value is in the way it teaches us the importance of taking care of our own world. After all, no matter how far we travel in the future, Earth will always be our home planet, and if we don't take good care of our home, no one else will."[32]

A short NASA video, "Benefits for Humanity: Water for the World," provides an excellent pairing with Bennett's book by presenting a concrete example of how working in space helps us take good care of our home planet and its people.[33] The Environmental Control and Life Support System (ECLSS) was developed to provide ISS astronauts with clean water for drinking, cooking, and hygiene. It has since been adapted to provide communities around the world with access to clean water. The video takes students to the village of Tres Pecos in Mexico to learn how NASA's solar-powered technology provides free drinking water for villagers and improves their health and the local economy. A campus assistant at a local school says "there were lots of children with parasites and stomach bugs before the purified water came. Now the students are healthier. . . . Children come with their cups at recess, and they get the water themselves."

Following the video, students are encouraged to reflect upon and discuss what their day would be like without access to clean water at home or in school. As they think about what such an experience might be like, they empathize with their Mexican peers while deepening their appreciation of how life on board the ISS helps solve problems on Earth.

"A Day without Satellites" is a multimedia initiative designed to foster students' appreciation of their dependence on satellites and educate others about the importance of protecting the outer space environment. It can be eye-opening for students to imagine what a day would be like without accurate weather forecasts to help them plan their day or without GPS to find their way in unfamiliar territory. Students work together in groups to prepare a presentation about the importance of satellites and subsequently create and perform skits dramatizing what a day would be like without satellites. Prior to writing their skits, students engage in classroom discussion to consider the many areas of daily life where we depend on satellites and what life would be like without them. Students research specific satellites to learn about the information they provide and why it is important; prepare presentations to share what they have learned with others; and finally, work in groups to write and perform their skits.

A book by David Baker and Heather Kissock titled *Satellites* serves as a literary lens to shape this initiative and foster student knowledge about the variety of roles that satellites play in our lives. The book is from the All about Space Science AV[2] media-enhanced books series which integrates digital content into the reading of the hard-copy book. Books in the series all contain informative video and audio clips, along with web links, slide shows, activities, and hands-on experiments for further research and study.

Students learn about the history and science of space exploration, the different parts of satellites, and aerospace careers. They learn how satellites are essential for weather forecasting, monitoring climate change, identifying sources of natural resources, and detecting the health of Earth's forests. They also learn how satellites identify sources of pollution; aid in urban planning; provide communication vehicles for television, telephone, and internet; and assist in navigation. The mission of the International Space Station—the largest satellite orbiting Earth—is also described in this book.

Satellites serves as a springboard for a classroom discussion that prepares students to undertake the initiative by focusing on the basic question: What are the many ways we depend on satellites? Students share what they have already learned about satellites and what they would like to learn. They are asked to share ways that people rely on satellites in their daily lives, often without realizing it. As the discussion evolves, teachers help students generate a list of the five major areas in which we depend on satellites: weather forecasting, communications, navigation, safety/emergency management, and environmental monitoring/health. It is important for students to understand that the images and data provided by satellites give us an overall perspective from space not possible to obtain on Earth, and that satellites monitor changes on Earth over time—two major reasons why satellite images and data are so valuable.

Once the main areas of our dependence on satellites have been identified, students view a brief video by space environmentalist Moriba Jah: "What If Every Satellite Suddenly Disappeared?"[34] The animated video stimulates student thinking about what a day without satellites would be like. There is no more international programming on television. Air traffic controllers have a difficult time preventing plane crashes. Stoplights and traffic control systems stop synchronizing. Without satellite-based timestamps, the world economy shuts down. The disruption of supply chains for food and medicine forces people to survive on what is locally available. The video concludes with a warning: "The space kilometers above our heads is like our forests, the ocean's biodiversity and clean air. If we don't treat it as a finite resource, we may wake up one day to find we no longer have it at all."

Following the video and with the teacher's guidance, students identify everyday activities they—or people they know—engage in under each of five areas of satellite dependence. Students then pose questions about how these activities would be affected if satellites were to disappear. The five areas with accompanying activities and questions generated by students might look something like this, with additional resources included for each area.

WEATHER FORECASTING

Weather satellites help us observe the Earth from space to predict weather patterns and make accurate weather forecasts. Information from weather satellites provides early warnings of severe storms, helping people prepare and stay safe. Activities: Dressing appropriately for the weather, planning outings with friends, and preparing for severe weather. Questions: How shall I dress for school today? Will there be enough snow to go sledding this weekend?

When will the predicted hurricane hit our city? How can we best prepare? Resource: NOAA SciJinks video titled "Meet a GOES-R Series Weather Satellite."[35]

COMMUNICATIONS

Communications satellites make it possible to view live television broadcasts from around the world, help people find where they need to go, and provide essential communications data in times of natural disasters. Activities: Watching TV broadcasts from around the world and accessing satellite internet. Questions: Will I be able to watch the Olympics without satellite TV? Will I be able to hear live news reports broadcast from another part of the world? Will my cousins—who live in a remote rural area without cable—lose their satellite internet? Resource: PBS LearningMedia video titled "How Do Satellites Work?"[36]

NAVIGATION

GPS satellites circle the Earth twice a day, using signals to determine a user's location and give directions to other places. Activities: Finding the way to an unfamiliar location and tracking a flight. Questions: How will we find our way to an unfamiliar location without GPS? Will I be able to track my uncle's flight next week when he flies to Europe on a business trip? Resource: NASA Space Place article titled "How Does GPS Work?"[37]

SAFETY/EMERGENCY MANAGEMENT

Satellite data and images promote safety in innumerable ways, such as tracking wildfires and measuring their temperatures to help firefighters predict the path of the fire and pinpoint the areas affected. Activities: Locating someone quickly who needs help, deploying first responders quickly and efficiently, and avoiding volcanic ash when flying. Questions: If someone is having a medical emergency, will the ambulance driver find the house quickly without GPS? If a forest fire is spreading, will the firefighters know where the need is greatest? Will the pilot of a plane flying near a volcano be able to avoid flying through clouds of volcanic ash? Resources: NOAA article titled "How Do Satellites Help Save Lives?"[38] and NASA animation titled "NASA Tracks Volcanic Ash with Satellites."[39]

ENVIRONMENTAL MONITORING/HEALTH

Satellite images and measurements are important for solving climate change because they play a key role in monitoring air, water, plant, and animal life. Satellites also monitor the location and severity of environmental disasters, such as oil spills, as well as tracking and mapping the movements of endangered species. Activities: Monitoring air and water pollution, determining the location and severity of environmental disasters, and studying climate change. Questions: Will scientists be able to effectively monitor and address climate change without Earth-observing satellites to provide measurements of Earth's ozone layer, air temperatures, ocean temperatures, sea levels, ice levels, greenhouse gas emissions, atmospheric gases, and deforestation? Will scientists know the extent of air and water pollution in a particular area so that people can be warned, and the problem solved? If an oil pipeline bursts in the ocean, will it be possible to monitor the situation and direct cleanup crews where to go? Resources: NASA website titled "Climate Kids–NASA's Eyes on the Earth"[40] and NASA website/app titled "Images of Change."[41]

After students have identified and discussed the major areas of satellite dependence, listed everyday activities under each area, and posed questions about how these activities might be impacted without satellite information, the class is divided into five small groups, with each group assigned an area (i.e., weather forecasting, communications, navigation, safety/emergency management, and environmental monitoring/health). Working together, group members determine what information satellites from their assigned area provide and why this information is important. Each group member chooses one satellite from that area to research and prepares a slide (utilizing PowerPoint or other presentation software) that includes a picture of the satellite, its launch date, and its altitude; a description of what the satellite does; and why the information it provides is important. Group members then combine their slides to create their group presentation and each group shares its presentations with the class. When the presentations are completed, the teacher reviews with students what they have learned. As a result of completing, presenting, and viewing the presentations, students are able to list the five major areas of satellite dependence, describe the information satellites from each area provide, and explain why this information is important. Students now have the background knowledge they need to write their skits.

Next, students work in small groups to write a skit about what a day without satellites might be like. As they brainstorm ideas for their skits, the teacher encourages students to focus on how satellites inform good decision-making, and how their everyday decisions would be affected by the absence of satellites. Students create scenes for their skits based on the major areas of satellite dependence their class has identified and the questions they have generated regarding how the loss of satellite data and images would affect everyday actions, activities, and events. As students think together about what a day in their lives might be like without satellites, they come up with thoughtful and creative scenarios.

Once students have finished writing their skits, they add simple props, costumes, and music as desired and perform them for their class. The skits may also be shared with other audiences—such as other classes, a school-wide community meeting, or a STEM-based parent night—with the slide presentations serving as a helpful way to introduce and frame the skits. The presentations and skits serve as powerful vehicles for educating others about the importance of protecting the outer space environment.

Students may wish to conclude their skits with a performance of an inspirational song entitled, "ISS (Is Somebody Singing)," recorded in 2013. The first space-to-Earth musical collaboration in history, ISS Commander Chris Hadfield sang in space and was joined by the Barenaked Ladies and the Wexford Gleeks singing on Earth.[42] The lyrics, posted online, underscore the close relationship between life in space and life on Earth.[43] Students enjoy singing the song and it helps listeners appreciate the value of orbital space.

The young readers' version of astronaut Scott Kelly's autobiography, *My Journey to the Stars*, serves as a literary lens for the second essential principle of space literacy: Human activity in space creates an orbiting junkyard of debris that threatens spacecraft and astronauts. Kelly writes about his childhood experiences in West Orange, New Jersey, that led him to become an astronaut, including his struggles in school and his difficult home life. He was close to his twin brother, Mark, who also became an astronaut. Kelly writes:

> At home, every day was different. Our parents didn't always get along. We never knew when a fight might start. It was scary when they argued. Mark and I would hide in our room. Watching our parents fight turned Mark and me into peacemakers. We can get along with almost anyone. We learned to stay calm in the toughest of times. This has helped us each lead our own crews in space. You never know when your problems can become your strengths.[44]

When Kelly's mother decided she wanted to become a police officer, his father—who was on the local police force—set up a training course for her in their backyard. Kelly writes:

> The hardest part was the wall Mom had to climb. Every day, she tried to climb over that wall. At first, she couldn't even touch the top. It took Mom a long time, but she finally did it. Soon after, Mom passed the test and became one of the first female cops in New Jersey. Her success showed Mark and me that if we had a plan, with small steps we could turn a big dream into something real.[45]

In high school, Kelly joined the volunteer ambulance unit and discovered that he enjoyed helping people. Visiting a bookstore, he came across a book by Tom Wolfe about the original NASA astronauts entitled *The Right Stuff*. It changed his life and inspired him to become an astronaut. This aspiration gave him a goal to work toward, and he developed a step-by-step plan. He worked hard at his studies, graduated from college, joined the US Navy, and learned how to fly. His first experience in space was flying on the space shuttle *Discovery* to fix the Hubble telescope. He subsequently flew on the space shuttle *Endeavor* and then spent a year in space on board the ISS. He discusses the joys and challenges of life in space, including the time the ISS was almost hit by space debris. The mix of vivid illustrations and photographs of Kelly's family that accompany his narrative help students empathize with him through the challenges he faced in his youth and the dangers he encountered in space. Students are encouraged to aim high and persevere through difficulties.

After reading the book together, students share what they think life in space must be like. They address this question: "What do you think is the most exciting part of being an astronaut, and the most dangerous part?" As the discussion develops, an important point to emphasize is that space debris endangers astronauts and spacecraft.

A "NASA Knows" article titled "What Is Orbital Debris?" and accompanying graphic provide additional information for students about the pollution of the orbital space environment and the importance of cleaning up space debris to keep astronauts and spacecraft safe.[46] The NASA Orbital Debris Program Office online photo gallery and space debris animation help students visualize how debris orbits Earth and how crowded space is becoming.[47]

As one student told his father after viewing a NASA space debris graphic, "If you think my room is messy, you should see what they're doing to outer space." His comment inspired a scenario that helps students relate the space debris problem to everyday life. The scenario is as follows: "Imagine that all the things you ever owned, including your broken toys and old clothes, were never given away or thrown out, but were allowed to pile up around your room. What would your room look like today? How messy do you think it would be? Do you think you might trip or fall over something and hurt yourself?" Through class discussion, students clearly recognize that just as they are expected to pick up their room and throw out their trash at home, the space environment must also be kept clean so that space junk does not pile up and harm astronauts or destroy satellites.

A short, gripping PBS video titled "A Year in Space—Houston, We Have a Problem: Dodging Space Debris"[48] provides an excellent pairing with Kelly's book and NASA resources. Students view Astronaut Scott Kelly's harrowing experience in July 2015 when the ISS dodged a piece of debris from an obsolete Russian weather satellite that could have ended his trip. Students watch as mission control at the Johnson Space Center notifies Kelly of a late notice "conjunction" that sends him and two Russian cosmonauts into the Soyuz capsule. They see the astronauts prepare to make a quick getaway before receiving word from Moscow mission control that the danger has passed. Students empathize with Kelly and the cosmonauts who face danger in space and are impressed by how calm and matter of fact they are.

The initiative for the second principle of space literacy is a narrative-writing assignment that students undertake after they have read and discussed Kelly's book and viewed the video. Following the video, the teacher asks students to discuss how they would react if they were in Kelly's position and received a "late notice conjunction" warning. After discussion, students are assigned the following writing prompt: Imagine that you are the commander on board the International Space Station. Suddenly, Mission Control notifies you that you must evacuate immediately, due to the danger of a major collision with a piece of space junk. How do you feel? What do you do next? What happens? What do you learn from this experience? Include details from the NASA article titled "What Is Orbital Debris?" in your narrative.

Students are encouraged to use clear event sequences, descriptive details, and dialogue to portray experiences and events, and to show the responses of characters to situations. Once they have written, edited, revised, and finalized their narratives based on peer and teacher feedback, the narratives become effective vehicles for students to use in advocating for the protection of the space environment to keep astronauts and spacecraft safe. By publicizing the narratives in a class newsletter or schoolwide publication, or by sharing them in a community meeting, they become effective vehicles for students to use in educating a wider audience about the debris problem and advocating for the protection of the orbital space environment.

Geckos, a book from the Nature's Children series by Katie Marsico, serves as a literary lens for principle 3: Solving the debris problem requires new technologies and citizen advocacy for new rules and international agreements. The book begins with a "Fact File," a helpful one-page overview of gecko characteristics. The five chapters that follow provide information about the various gecko species, including their shapes and sizes, adaptations that help them survive in the wild, unique physical characteristics, how they feed, their life cycle, everyday activities and behavior, and how humans are affecting their chance for survival. The final chapter addresses how human activities have led certain species to be listed as threatened or endangered. The book concludes with a habitat map that shows where geckos are located. Elementary-age students find the book especially appealing because of the colorful, vivid photographs that alternate with pages of well-organized, content-rich text.

Students also learn about the adaptations on geckos' feet that enable them to readily climb smooth surfaces. Marsico writes:

> Geckos' feet are an example of the amazing adaptations that help them survive in the wild. Most gecko species can run up smooth, vertical surfaces with ease. Many animals would find this type of movement challenging. Geckos can do this because they have millions of tiny hairs called setae on the bottoms of their feet. The setae are split into hundreds of tips known as septulae. Together, the setae and septulae create a grip. The result is that geckos' feet have a built-in adhesive that allows the reptiles to cling to whatever surface they are walking across.[49]

Focusing on geckos, students learn about biomimicry—the use of animal characteristics to design solutions to problems. A video titled "Crazy Engineering: Gecko Gripper" from NASA's Jet Propulsion Laboratory explains how geckos' feet have inspired NASA engineers to devise a promising solution to removing debris from orbit.[50] Students learn how synthetic gecko adhesives function on grippers that grapple and manipulate objects in NASA's zero gravity aircraft called the "vomit comet." They also learn about van der Waals forces, the science behind geckos' feet that make them "nature's most amazing climbers." The idea that biomimicry can be useful in solving a problem in space serves to reinforce the close connection between Earth and space for students.

Designing a space junk cleanup robot is the initiative for the third principle of space literacy. After the class views the video, the teacher poses this question: How is biomimicry helping to solve the space debris problem, and how might it continue to be used in the future? Students are encouraged to share ideas with their classmates about the characteristics of other plants and animals, in addition to the gecko, and how these characteristics could be useful in designing space debris solutions.

Next, students are assigned the following writing prompt: Think about your favorite animal or plant. What characteristics does it have? What does it do especially well? Is it good at trapping prey with long tentacles or sharp claws? Does it run fast? What new technology to trap and remove space debris could be inspired by the special characteristics of your plant or animal?

Students are given the opportunity to share their ideas with the class. Then they work individually or in small groups to design a space junk cleanup robot utilizing biomimicry, incorporating characteristics of animals or plants. After designing their robot, students write an informative/explanatory paragraph beneath their picture, describing the qualities of their robot and how these qualities might be helpful in cleaning up space debris.

Students share their robot designs and accompanying paragraphs with a larger audience by displaying them in the classroom, school hallways, or the cafeteria. The student work becomes a way to stimulate exploration and discussion regarding the space debris problem and potential solutions. Students also can share their designs at a schoolwide community meeting or curricular event, include them in a class newsletter, or show and explain them to other classes. A local community member who works in the aerospace industry might be enlisted to hear the student presentations and offer feedback.

Students also could write a one-page advocacy letter to their US Representatives and Senators. It is important for them to let their legislators know they are educated about the topic of space debris. They want to express their opinion about this important issue and urge Congress to take action that will protect the outer space environment.

Students discuss several questions as a prewriting activity: Why is it important to protect the space environment from debris? What problems are caused by debris? How serious is the threat posed by debris? What are some of the possible solutions? What should the United States do to clean up the orbiting junkyard in space? Students share their responses to these questions in classroom discussion and then write their letters to Congress. They may also wish to adapt their letters for submission to a local newspaper or other news media.

In his adult memoir, *Endurance,* Scott Kelly recounts his call to author Tom Wolfe from the ISS:

> I've been thinking about the whole arc of my life that brought me here, and I always think about what it meant to me to read *The Right Stuff* as a young man. I feel certain I wouldn't have done any of the things I have if I hadn't read that book—if Tom Wolfe hadn't written it. On a quiet Saturday afternoon, I call Tom Wolfe to thank him. He sounds truly amazed to hear from me. I tell him we're passing over the Indian Ocean, how fast we're going, how our communications system works. We talk about books and about New York and about what I plan to do first when I get back (jump into my swimming pool). We agree to have lunch when I'm back on Earth, and that's now one of the things I'm looking forward to most.[51]

Several months before his call to Wolfe, Kelly and his crew experienced what he describes as a "potentially disastrous scenario" with space debris. In the event of a collision, he later reflected, the astronauts would have been "blasted in a million directions, as diffused atoms, all in the space of a millisecond. Our neurological systems would not even have had time to process the incoming data into conscious thought." He wasn't sure whether the thought comforted him or disturbed him.[52]

It took a visionary scientist inspired by *War of the Worlds*, Robert Goddard, to open the door to the space age. The question now is whether the international community will act in time to prevent the door from closing as key orbits become congested with debris. It is going to take the best efforts of visionary public servants, such as the United Nations' Simonetta di Pippo, and public-spirited citizens to keep the door open.

Teachers are well-positioned to play a role in this work by educating space-age environmentalists who have the knowledge and motivation needed to address the dual threats of space debris and climate change. Authors such as Jeffrey Bennett are important partners in this educational venture. In *Max Goes to the Space Station*, he appeals to the highest aspirations and best qualities in every child:

> We are living at a turning point in human history. Today, we face many serious global problems, including problems caused by war and hatred, and problems that we create through pollution and global warming. Our future depends on solving these problems, and our best chance of solving them will come if children everywhere are inspired to dream, to work hard, and to build a better future. There's no better source of this inspiration than space exploration, which reminds us that everything is possible.[53]

Sometimes educators are curious to know what stirred our interest in the space debris issue. In 2007, we read a summary of the Fourth Assessment Report of the Intergovernmental Panel on Climate Change and were struck by the extent to which it relied on satellite data for its conclusions.[54] We thought we should learn more about the spacecraft that provide this indispensable information. We discovered, for example, that scientists had used data collected between 1996 and 2005 by the European Space Agency's (ESA) remote sensing satellites ERS-1 and ERS-2, ESA's Envisat Advanced Synthetic Aperture Radar (ASAR), and the Canadian Space Agency's Radarsat-1 to reach their conclusion that Greenland glaciers were melting into the sea twice as fast as previously believed.[55]

We also came across a small body of literature on the space debris problem and the threat it poses to satellites we depend upon to fulfill our responsibility to care for Earth. As we educated ourselves about this issue, we realized we had to expand our environmental perspective and activities to encompass orbital space as well as Earth. We encourage educators to introduce the new topic of space literacy to students as an essential complement to climate literacy by test-flying concepts and materials from this chapter.

NOTES

1.Arthur Scott Geddes, "UN Space Chief Calls for Mass Action to Save Earth and Cosmos," *The National*, September 29, 2020, https://www.thenationalnews.com /uae/environment/un-space-chief-calls-for-mass-action-to-save-earth-and-cosmos-1 .1085174.

2. Geddes.

3. European Space Agency, "Space Debris by the Numbers," last modified May 10, 2022, https://www.esa.int/Safety_Security/Space_Debris/Space_debris_by_the _numbers.

4. Lee Billings, "Under Trump, NASA May Turn a Blind Eye to Climate Change," *Scientific American*, November 23, 2016, https://www.scientificamerican.com/article /under-trump-nasa-may-turn-a-blind-eye-to-climate-change/.

5. European Space Agency, "Types of Orbits," March 30, 2020, https://www.esa.int /Enabling_Support/Space_Transportation/Types_of_orbits.

6. Brian C. Weeden and Tiffany Chow, "Developing a Framework and Potential Policies for Space Sustainability Based on Sustainable Management of Common-Pool Resources," IAC-11.E3.4.3, 62nd International Astronautical Congress, Cape Town, South Africa, October 2–7, 2011, 4–5, https://swfound.org/media/50234/iac-11.e3.4 .3-paper.pdf.

7. European Space Agency, "About Space Debris," accessed June 10, 2022, https: //www.esa.int/Safety_Security/Space_Debris/About_space_debris.

8. European Space Agency, "Space Debris by the Numbers," last modified May 10, 2022, https://www.esa.int/Safety_Security/Space_Debris/Space_debris_by_the _numbers.

9. European Space Agency, "Space Debris by the Numbers," last modified May 10, 2022, https://www.esa.int/Safety_Security/Space_Debris/Space_debris_by_the _numbers.

10. NASA Headquarters Library, "Space Debris," last modified July 1, 2019, https: //www.nasa.gov/centers/hq/library/find/bibliographies/space_debris.

11. NASA Office of Inspector General, "NASA's Efforts to Mitigate the Risks Posed by Orbital Debris," Report No. IG-21–011, January 27, 2021, 2, https://oig .nasa.gov/docs/IG-21-011.pdf.

12. NASA, "Micrometeoroids and Orbital Debris (MMOD)," last modified August 6, 2017, https://www.nasa.gov/centers/wstf/site_tour/remote_hypervelocity _test_laboratory/micrometeoroid_and_orbital_debris.html.

13. NASA Office of Inspector General, 14–15.

14. Sandra Erwin, "U.S. Military Keeps Sharp Eyes on Orbit as Congestion Grows," *Space News*, November 3, 2020, https://spacenews.com/u-s-military-keeps-sharp -eyes-on-orbit-as-congestion-grows/.

15. Scott Kelly, *Endurance: A Year in Space, A Lifetime of Discovery* (New York: Alfred A. Knopf, 2017), 179.

16. Martin Fackler, "Space's Trash Collector? A Japanese Entrepreneur Wants the Job, *New York Times*, November 28, 2016, https://www.nytimes.com/2016/11/28/ science/space-junk-astroscale.html.

17. Charles Q. Choi, "Gecko-Inspired Robot Could Snag Space Junk," June 28, 2017, video, 3:21, https://www.space.com/37335-robotic-gecko-gripper-microgravity-space-junk.html.

18. James Clay Moltz, *Crowded Orbits: Conflict and Cooperation in Space* (New York: Columbia University Press, 2014), 189.

19. NASA Office of Inspector General, 31.

20. NASA Office of Inspector General, 31.

21. Union of Concerned Scientists, "Climate Solutions," accessed June 9, 2022, https://www.ucsusa.org/climate/solutions.

22. Rebecca L. Young, *Confronting Climate Crises through Education* (Lanham: Lexington Books, 2018), 4.

23. Daphne Minner and Jennifer Klein, *A Case for Advancing an Environmental Literacy Plan in Massachusetts: Phase I—a Summary of the Commonwealth's Environment and Education Landscape* (Massachusetts Environmental Education Society, 2016), 16, http://massmees.org/wp-content/uploads/2017/03/MassELP-Phase-I-Summary-FINAL-REV-March2017.pdf.

24. Lee Billings, "War in Space May Be Closer Than Ever," *Scientific American*, August 10, 2015, https://www.scientificamerican.com/article/war-in-space-may-be-closer-than-ever/.

25. NASA, "Dr. Robert H. Goddard, American Rocketry Pioneer," https://www.nasa.gov/centers/goddard/about/history/dr_goddard.html.

26. Charles Bolden, "The Stuff of Goddard's Dreams: Goddard's Legacy & NASA's Journey to Mars," February 9, 2016, 2, https://www.nasa.gov/sites/default/files/atoms/files/bolden_goddard_2016.pdf.

27. Jamil Zaki, *The War for Kindness: Building Empathy in a Fractured World* (New York: Crown, 2019), 76.

28. Zaki, 81.

29. Story Time from Space, "Max Goes to the Space Station: A Science Adventure with Max the Dog," 2014, video, 16:52, https://storytimefromspace.com/stories/max-goes-to-the-international-space-station/.

30. Jeffrey Bennett and Michael Carroll, *Max Goes to the Space Station: A Science Adventure with Max the Dog* (Boulder: Big Kid Science, 2014), 24.

31. Bennett and Carroll, 21.

32. Bennett and Carroll, 29.

33. NASA, "Benefits for Humanity: Water for the World," last modified April 17, 2019, video, 3:31, https://www.nasa.gov/content/benefits-for-humanity-water-for-the-world.

34. Moriba Jah, "What If Every Satellite Suddenly Disappeared?," filmed February 2021, TED-Ed video, 4:58, https://www.ted.com/talks/moriba_jah_what_if_every_satellite_suddenly_disappeared?language=en.

35. National Oceanic and Atmospheric Administration SciJinks, "Meet a GOES-R Series Weather Satellite," accessed December 21, 2021, video, 1:53, https://scijinks.gov/goes-r/.

36. PBS Learning Media, "How Do Satellites Work?," Ready-Jet-Go Collection video, accessed December 21, 2021, video, 0:59, https://mass.pbslearningmedia.org /resource/ready-get-go-how-do-satellites-work/how-do-satellites-work-ready-jet-go/.

37. NASA Space Place, "How Does GPS Work?," last modified June 27, 2019, https://spaceplace.nasa.gov/gps/en/.

38. National Oceanic and Atmospheric Administration SciJinks, "How Do Satellites Help Save Lives?," accessed December 21, 2021, https://scijinks.gov/sarsat/.

39. NASA Goddard, "NASA Tracks Volcanic Ash with Satellites," May 12, 2016, video, 1:36, https://www.youtube.com/watch?v=zAXvSoo3F8A&t=83s.

40. NASA Climate Kids, last modified October 19, 2021, https://climatekids.nasa .gov/.

41. NASA Global Climate Change, "Images of Change," accessed December 21, 2021, https://climate.nasa.gov/images-of-change/.

42. Chris Hadfield and Ed Robertson, "I.S.S. (Is Somebody Singing)," February 12, 2013, music video, 5:28, https://www.youtube.com/watch?v=AvAnfi8WpVE.

43. Chris Hadfield and Ed Robertson, "I.S.S. (Is Somebody Singing)," lyrics, February 8, 2013, https://genius.com/Chris-hadfield-iss-is-somebody-singing-lyrics.

44. Scott Kelly, *My Journey to the Stars* (New York: Crown Books for Young Readers, 2017), unnumbered pages.

45. Kelly, unnumbered pages.

46. NASA, "What Is Orbital Debris?," NASA Knows! (Grades K-4) Series, last modified June 7, 2021, https://www.nasa.gov/audience/forstudents/k-4/stories/nasa -knows/what-is-orbital-debris-k4.html.

47. NASA Orbital Debris Program Office, "Photo Gallery," January 1, 2019, animation video, 2:20, https://orbitaldebris.jsc.nasa.gov/photo-gallery/.

48. PBS, "A Year in Space—Houston, We Have a Problem: Dodging Space Debris," March 2, 2016, video, 1:59, https://www.pbs.org/video/year-space-houston -we-have-problem-dodging-space-debris/.

49. Katie Marsico, *Geckos*, Nature's Children (New York: Scholastic Children's Press, 2014), 13.

50. NASA Jet Propulsion Laboratory, "Crazy Engineering: Gecko Gripper," August 12, 2015, video, 3:51, https://www.youtube.com/watch?v=6zasTmmR95E.

51. Kelly, *Endurance*, 350.

52. Kelly, 180–181.

53. Bennett and Carroll, 27.

54. IPCC, 2007: Summary for Policymakers. In: Climate Change 2007: The Physical Science Basis. Contribution of Working Group I to the Fourth Assessment Report of the Intergovernmental Panel on Climate Change [Solomon, S., D. Qin, M. Manning, Z. Chen, M. Marquis, K.B. Averyt, M. Tignor and H.L. Miller (eds.)]. Cambridge University Press, Cambridge, United Kingdom and New York, NY, USA, https://www.ipcc.ch/site/assets/uploads/2018/02/ar4-wg1-spm-1.pdf.

55. European Space Agency, "Satellite Data Yields Major Results in Greenland Glaciers Study," February 21, 2006, https://www.esa.int/Applications/Observing_the _Earth/Satellite_data_yields_major_results_in_Greenland_glaciers_study.

BIBLIOGRAPHY

Baker, David, and Heather Kissock. *Satellites.* All About Space Science. New York: AV² by Weigl, 2018.

Bennett, Jeffrey, and Michael Carroll. *Max Goes to the Space Station: A Science Adventure with Max the Dog.* Boulder: Big Kid Science, 2014.

Billings, Lee. "Under Trump, NASA May Turn a Blind Eye to Climate Change." *Scientific American,* November 23, 2016. https://www.scientificamerican.com/article/under-trump-nasa-may-turn-a-blind-eye-to-climate-change/.

———. "War in Space May Be Closer Than Ever." *Scientific American*, August 10, 2015. https://www.scientificamerican.com/article/war-in-space-may-be-closer-than-ever/.

Bolden, Charles. "The Stuff of Goddard's Dreams: Goddard's Legacy & NASA's Journey to Mars." Remarks, Robert H. Goddard Memorial Symposium, Greenbelt, MD. February 9, 2016. https://www.nasa.gov/sites/default/files/atoms/files/bolden_goddard_2016.pdf.

Choi, Charles Q. "Gecko-Inspired Robot Could Snag Space Junk." June 28, 2017. Video, 3:21. https://www.space.com/37335-robotic-gecko-gripper-microgravity-space-junk.html.

Erwin, Sandra. "U.S. Military Keeps Sharp Eyes on Orbit as Congestion Grows." *Space News*, November 3, 2020. https://spacenews.com/u-s-military-keeps-sharp-eyes-on-orbit-as-congestion-grows/.

European Space Agency. "About Space Debris." Accessed June 10, 2022. https://www.esa.int/Safety_Security/Space_Debris/About_space_debris.

———. "Satellite Data Yields Major Results in Greenland Glaciers Study." February 21, 2006. https://www.esa.int/Applications/Observing_the_Earth/Satellite_data_yields_major_results_in_Greenland_glaciers_study.

———. "Space Debris by the Numbers." Last modified May 10, 2022. https://www.esa.int/Safety_Security/Space_Debris/Space_debris_by_the_numbers.

———. "Types of Orbits." March 30, 2020. https://www.esa.int/Enabling_Support/Space_Transportation/Types_of_orbits.

Fackler, Martin. "Space's Trash Collector? A Japanese Entrepreneur Wants the Job." *New York Times*, November 28, 2016. https://www.nytimes.com/2016/11/28/science/space-junk-astroscale.html.

Hadfield, Chris. "I.S.S. (Is Somebody Singing)." February 8, 2013. Song lyrics. https://genius.com/Chris-hadfield-iss-is-is-somebody-singing-lyrics.

Hadfield, Chris, and Barenaked Ladies. "I.S.S. (Is Somebody Singing)." February 12, 2013. Music video, 5:28. https://www.youtube.com/watch?v=AvAnfi8WpVE.

IPCC, 2007: Summary for Policymakers. In: Climate Change 2007: The Physical Science Basis. Contribution of Working Group I to the Fourth Assessment Report of the Intergovernmental Panel on Climate Change [Solomon, S., D. Qin, M. Manning, Z. Chen, M. Marquis, K.B. Averyt, M. Tignor and H.L. Miller (eds.)]. Cambridge, UK, and New York: Cambridge University Press. https://www.ipcc.ch/site/assets/uploads/2018/02/ar4-wg1-spm-1.pdf.

Jah, Moriba. "What If Every Satellite Suddenly Disappeared?" Filmed February 2021, TED-Ed video. https://www.ted.com/talks/moriba_jah_what_if_every_satellite_suddenly_disappeared?language=en.

Kelly, Scott. *My Journey to the Stars.* Toronto: Crown Books for Young Readers, 2017.

Kelly, Scott, with Margaret Lazarus Dean. *Endurance: A Year in Space, A Lifetime of Discovery.* New York: Knopf, 2017.

Marsico, Katie. *Geckos.* Nature's Children. New York: Scholastic Children's Press, 2014.

Minner, Daphne, and Jennifer Klein. *A Case for Advancing an Environmental Literacy Plan in Massachusetts: Phase I—a Summary of the Commonwealth's Environment and Education Landscape.* Massachusetts Environmental Education Society, 2016. http://massmees.org/wp-content/uploads/2017/03/MassELP-Phase-I-Summary-FINAL-REV-March2017.pdf.

NASA. "Benefits for Humanity: Water for the World." November 20, 2013. Article and video, 3:31. https://www.nasa.gov/content/benefits-for-humanity-water-for-the-world.

———. "Dr. Robert H. Goddard, American Rocketry Pioneer." Last updated August 3, 2017. https://www.nasa.gov/centers/goddard/about/history/dr_goddard.html.

———. "Micrometeoroids and Orbital Debris (MMOD)." Last updated August 6, 2017. https://www.nasa.gov/centers/wstf/site_tour/remote_hypervelocity_test_laboratory/micrometeoroid_and_orbital_debris.html.

———. "What Is Orbital Debris?" *NASA Knows! (Grades K–4) series,* June 8, 2010. Last updated June 7, 2021. https://www.nasa.gov/audience/forstudents/k-4/stories/nasa-knows/what-is-orbital-debris-k4.html.

NASA Climate Kids. Last updated October 19, 2021. https://climatekids.nasa.gov/.

NASA Global Climate Change. "Images of Change." Accessed December 21, 2021. https://climate.nasa.gov/images-of-change/.

NASA Goddard. "NASA Tracks Volcanic Ash with Satellites." May 12, 2016. Video, 1:36. https://www.youtube.com/watch?v=zAXvSoo3F8A&t=83s.

NASA Headquarters Library. "Space Debris." Last modified July 1, 2019. https://www.nasa.gov/centers/hq/library/find/bibliographies/space_debris.

NASA Jet Propulsion Laboratory. "Crazy Engineering: Gecko Gripper." August 12, 2015. Video, 3:51. https://www.youtube.com/watch?v=6zasTmmR95E.

NASA Office of Inspector General. "NASA's Efforts to Mitigate the Risks Posed by Orbital Debris," Report No. IG-21–011. January 27, 2021. https://oig.nasa.gov/docs/IG-21-011.pdf.

NASA Orbital Debris Program Office. "Photo Gallery." January 1, 2019. Animation video, 2:20. https://orbitaldebris.jsc.nasa.gov/photo-gallery/.

NASA Space Place. "How Does GPS Work?" Last updated June 27, 2019. https://spaceplace.nasa.gov/gps/en/.

National Oceanic and Atmospheric Administration SciJinks. "How Do Satellites Help Save Lives?" Accessed December 21, 2021. https://scijinks.gov/sarsat/.

National Oceanic and Atmospheric Administration SciJinks. "Meet a GOES-R Series Weather Satellite." Accessed December 21, 2021. Video, 1:53. https://scijinks.gov /goes-r/

PBS. "A Year in Space—Houston, We Have a Problem: Dodging Space Debris." March 2, 2016. Video, 1:59. https://www.pbs.org/video/year-space-houston-we -have-problem-dodging-space-debris/.

PBS Learning Media. "How Do Satellites Work?" Ready-Jet-Go Collection video. Accessed December 21, 2021. Video, 0:59. https://mass.pbslearningmedia.org /resource/ready-get-go-how-do-satellites-work/how-do-satellites-work-ready-jet -go/.

Scott-Geddes, Arthur. "UN Space Chief Calls for Mass Action to Save Earth and Cosmos." *The National.* Updated September 29, 2020. https://www.thenation- alnews.com/uae/environment/un-space-chief-calls-for-mass-action-to-save-earth -and-cosmos-1.1085174.

Story Time from Space. *Max Goes to the Space Station: A Science Adventure with Max the Dog.* Video, 16:52. https://storytimefromspace.com/stories/max-goes-to -the-international-space-station/.

Union of Concerned Scientists. "Climate Solutions." https://www.ucsusa.org/climate /solutions.

Weeden, Brian C., and Tiffany Chow. "Developing a Framework and Potential Policies for Space Sustainability Based on Sustainable Management of Common-Pool Resources." Paper, 62nd International Astronautical Congress, Cape Town, South Africa, October 2–7, 2011. https://swfound.org/media/50234/iac-11.e3.4.3-paper .pdf.

Young, Rebecca L. *Confronting Climate Crises through Education.* Lanham, MD: Lexington Books, 2018.

Zaki, Jamil. *The War for Kindness: Building Empathy in a Fractured World.* New York: Crown, 2019.

Chapter Four

Teaching Environmental Respect to Young Learners

Video Games as Environmental Texts

Erden El

As environmental concerns are rising at an unprecedented pace, literary studies could not remain insensitive to this situation. Literary critics working in the field of ecocriticism aim to raise awareness in society by applying ecocritical theory to literary works. Deforestation, pollution and diseases caused by environmental abuses have increased. Since literature is a mirror of life, authors with environmental awareness began to reflect this situation. *Silent Spring* (1962) by Rachel Carson proves that books can be very useful tools for raising awareness. Carson's ground-breaking book explores the topic of DDT use and the dangers it poses. Her field research allowed her to demonstrate the hazards of DDT to other living beings, specifically the birds she observed. Though Carson faced heavy criticism by companies making money on DDT and advocates of DDT use, her book was so successful that it both caused the prohibition of DDT use and positioned Carson as a pioneer of the environmental movement. Linda Lear, the author of the introduction to the 1962 edition of *Silent Spring*, describes the hardships Carson endured:

In 1962, however, the multimillion-dollar industrial chemical industry was not about to allow a (. . .) female scientist (. . .) known only for her lyrical books on the sea, to undermine public confidence in its products or to question its integrity. It was clear to industry that Rachel Carson was a hysterical woman whose alarming view of the future could be ignored or, if necessary suppressed. She was a "bird and bunny lover," a woman who kept cats and was therefore clearly suspect. She was a romantic "spinster" who was simply overwrought about genetics. In short, Carson was a woman out of control. She had overstepped the bounds of her gender and her science.[1]

Despite all these humiliations, Carson resolutely initiated the banning of DDT by influencing and galvanizing the public thanks to her advocacy for awareness. It is of note for this collection of resources that Carson began *Silent Spring* with a fable. Its cautionary message about a "tomorrow" where no birdsong could be heard was a clear call to action for the public, who themselves could observe the environmental effects of DDT.

It is similarly valuable today to read a text by examining its subtexts from an ecocritical perspective. This perspective enables one to analyze the ways in which human action exploits natural forces. Contemporary humanities do not render the word text merely as a written text but include film and video games; the discourse of any text can be analyzed. Because the written text is the heart of ecocriticism and young people are very interested in video games, which makes the ecocritical approach more understandable for them, this chapter will examine both.

To increase the environmental awareness of young people who will build the future, this study addresses mainly but not only high school students and their teachers. Hoping that this approach will serve as a resource for those who may be new to ecopedagogy, the chapter begins with an introduction to ecocriticism and a brief analysis of *White Noise* (1984) by Don DeLillo as an example. The study will then introduce a curricular foundation for studying video games ideal for ESL/EFL classes in high schools. The method of this instructional approach is to apply ecocritical theory to literary works and video games simultaneously. It aims to be a guide for teachers who want to teach their students the concepts of ecocriticism. In addition, it aims to provide students with the ability to transfer theory to practice.

ECOPHOBIA OR DESIRE TO DOMINATE NATURE

I recommend that teachers mindfully inform their students that their aim is not to criticize religion while introducing this topic. It is the perspective of people that is criticized here; religious texts are merely examples. There is a common belief that human beings are in the form of God and are the purposeful, intelligent creations of God who have been sent to Earth with a mission to build civilization. These overestimated assumptions on human life are a result of how people want to see themselves. It is an indirect way of saying if God has control over the universe, his representatives should also have dominion. For human beings, the belief that the universe exists ultimately for them leads to abuses of nature and nonhuman beings:

And God said, let us make man in our image, after our likeness: and let them have dominion over the fish of the sea, and over the fowl of the air, and over the cattle, and over all the earth, and over every creeping thing that creepeth upon the earth. So God created man in his own image [. . .] and God blessed them, and God said unto them, be fruitful, and multiply, and replenish the earth, and subdue it: and have dominion over the fish of the sea, and over the fowl of the air, and over every living thing that moveth upon the earth.[2]

It is stated in Genesis those human beings should expand on earth, reshape and tame the world. The initial existence of the world is represented as unruly, dangerous, and uncanny, and it is essential to establish an order based on human hegemony in the world. This perception contemplates the world as hostile toward human beings, and human beings are summoned to rage war against the enemy. Such a perception of the earth can never overcome the distrust or animosity of nature, and it makes it impossible for those who follow this view to transcend anthropocentricism. This study does not intend to attack any religion. It is meant to illustrate a specific and influential literary example that human beings seek justification in religious texts for their environmentally hostile behaviors.

The desire to civilize nature exists in ancient legends too. "Gilgamesh" is a magnificent work that conveys the fear that ancient societies had of nature and natural phenomena. Gilgamesh is the tyrannical King of Uruk, and Enkidu is the newcomer in Uruk, who was "brought up by the animals of the wild."[3] Gilgamesh, the hero of the Babylonian legend, represents the modern man of that period, while Enkidu represents the wild nature with its satyr-like appearance. In this sense, nature/culture dichotomy, which is the argument that nature and culture are opposite concepts, dates far back in time. Gilgamesh, who brought civilization to the land he lived in, is praised for his power in the epic. But interestingly, Gilgamesh, "the perfect in strength"[4] who dominates nature, is praised on the one hand and represented as a cruel king on the other since the epic refers to Gilgamesh both as "magnificent and terrible."

It is evident in the text that there is both a fear and a reverence for Gilgamesh. The respect for Gilgamesh stems from fear. On the other hand, Endiku lives as a wild person in nature before being tamed by Gilgamesh. His death after being tamed also shows that he could not adapt to this new civilization.

ECOCRITICISM: A LITERARY SCHOOL WHICH
DEALS WITH THE ENVIRONMENT

The term *ecocriticism* was coined by William Rueckert in his essay "Literature and Ecology: An Experiment in Ecocriticism" (1978). Rueckert defines the term "ecocriticism" as "the application of ecology and ecological concepts to the study of literature."[5] Although this phrasing may sound too general, it is very significant because the study of the environment as a theme in literature has gained a definition thanks to Rueckert. Coining the term gathered the studies under a certain discipline and accelerated them. Rueckert's main argument was the distinction between "biocentrism" and "anthropocentrism."[6] According to Rueckert, "in ecology, man's tragic flaw is his anthropocentric (as opposed to biocentric) vision, and his compulsion to conquer, humanize, domesticate, violate, and exploit every natural thing."[7] By stating that, Rueckert called attention to what ecocritics have criticized: an attitude toward nature and the environment that is human-centered. The faulty mentality of human beings derives from the relentless effort to grow, tame and humanize the land. This topic later developed into the nature/culture dichotomy, which is a subject commonly debated in the field of ecocriticism.

In her book *The Ecocriticism Reader* (1996), which she co-edited with Harold Fromm, Cheryll Glotfelty developed the most generally recognized concept of the word *ecocriticism*. Ecocriticism is described by Glotfelty as "the study of the relationship between literature and the physical environment."[8] She also notes that "just as feminist criticism examines language and literature from a gender-conscious perspective, and Marxist criticism brings an awareness of modes of production and economic class to its reading of texts, ecocriticism takes an earth-centered approach to literary studies."[9] According to Glotfelty, the ecocritics' approach should be to address the following questions in their critical essays:

1. How is nature represented in this (work)?
2. What role does the physical setting play in the plot of this novel?
3. Are the values expressed in this play consistent with ecological wisdom?
4. How do our metaphors of the land influence the way we treat it?
5. How can we characterize nature writing as a genre?
6. In addition to race, class, and gender, should place become a new critical category?
7. Do men write differently about nature than women?
8. How has the concept of wilderness changed over time?
9. In what ways is the environmental crisis seeping into contemporary literature and popular culture?

10. What bearing does the science of ecology have on literary studies?
11. How is science itself open to literary analysis?[10]

The research questions formulated by Glotfelty are very beneficial for taking an ecocritical approach to analysis. An instructor can offer insight into ecocriticism, particularly in practice, by encouraging students to consider these questions as they interact with texts. Students will develop a broader insight and may redefine their interaction with nature by integrating these research questions and their responses to them in analyses of literary texts and other media.

The theory of ecocriticism, from its emergence in the 1960s to the present time, and its applications for studying literary texts, has been drawing attention to environmental issues. Literary texts have been ecocritical models in which the natural environment, biota and plants, water and air pollution, waste and human toxicity, are considered. How culture is rendered is an especially significant issue. Is culture seen as a way to conquer nature? Or is it a way to cooperate with nature? The representation of culture and misrepresentations of it make up the point of view of the text, which the ecocritic analyzes. Another matter of curiosity is whether the misrepresentation of nature is the idea of the writer or whether the characters symbolize misrepresentations of nature. It is also possible that the writer creates characters who defend the misinterpretation of nature to help readers make a commentary about it. False beliefs and human-centered perspectives observed in characters may also be fictionalized by the author to promote reader awareness. All these possibilities must be taken into account but rather than the author's intention, the focus should be on environmental factors present in the text.

The relationship between nature and human beings forms the basis of ecocriticism. How human beings evaluate nature and nonhuman beings, including perspectives toward them, is an element that ecocritics carefully examine. If this element, which can be named as cultural representation, reflects the desire of the entity called culture to dominate nature, this can be called misrepresentation. Like many ecocritics, I also do not accept the so-called nature/culture dichotomy. But unfortunately, such a distinction is made commonly. Although I do not accept this distinction, I reference it for a better enlightenment of the subject and choose to call it a "cultural misrepresentation." As Gomides mentions, "the field analyses and promotes works of art which raise moral questions about human interactions with nature, while also motivating audiences to live within a limit that will be binding over generations."[11]

In his work *Writing the Climate*, Richard Kerridge discusses the ecocritical working technique (1998). The ecocritic should, according to Kerridge, expose the depictions of nature in a literary text and reread the text from an ecocentric viewpoint:

The ecocritics want to track environmental ideas and representations wherever
they appear, to see more clearly a debate which seems to be taking place, often
part-concealed, in a great many cultural spaces. Most of all ecocriticism seeks to
evaluate texts and ideas in terms of their coherence and usefulness as responses
to environmental crisis.[12]

As Kerridge mentions, the discourse related to nature may not be explicit but
rather implied. It is the duty of the ecocritic to unearth the representations of
nature from the setting and/or the characters' dialogue or reflections.

Although the main concern of ecocriticism is nature, the scope of ecocriticism is not limited to nature. Ecocriticism strives for a positive change in the
perspective of human beings on nature. Estok argues that it would be wrong
to reduce ecocriticism to merely a study of nature since it is not

simply the study of Nature or natural things in literature; rather, it is any theory
that is committed to effecting change by analyzing the function—thematic,
artistic, social, historical, ideological, theoretical, or otherwise—of the natural
environment, or aspects of it, represented in documents (literary or other) that
contribute to material practices in material worlds.[13]

The material practices, as Estok mentions, are a significant dynamic of the
natural world. The material world and the nonhuman world are a part of life,
and ecocriticism strives to reveal this reality as well as dealing with the study
of nature.

Should the representation of nature be explicit so the ecocriticism can be
applied to the text? Should the text always be one that seeks to illuminate
environmental issues such as pollution or deforestation? Should the characters in a text be environmental activists? The answer to all these questions is
no. It is helpful if the works have these characteristics. However, it is not an
obligation, and, just as Slovic mentions, the text can even be indifferent to
nature.[14] Slovic states that ecocriticism is

the study of explicitly environmental texts by way of any scholarly approach or,
conversely, the scrutiny of ecological implications and human-nature relations
in any literary text, even texts that seem, at first glance, oblivious of the nonhuman world. In other words, any conceivable style of scholarship becomes a form
of ecocriticism if it's applied to certain kinds of literary works.[15]

As Slovic mentions, ecocriticism is not limited to environmentally conscious
works; any kind of literary texts can be studied in the light of ecocriticism.[16]
Even a nonliterary text can be a matter of ecocriticism—a documentary, a
film, or even a song can be the subject of ecocritical research.

To make this pedagogical approach more comprehensible and interesting for young learners, this chapter explores video games such as *Reset Earth* and *The Fate of the World*. As an illustrative reference for how to examine literary works from an ecocritical perspective, after explaining the theory, I will refer to Don DeLillo's *White Noise* (1984). After showing how the theory of ecocriticism can be practiced with this novel, I present an approach that demonstrates how educators can look at popular games from an ecocritical perspective.

AN EXAMPLE OF ECOCRITICAL ANALYSIS

This part of the work aims simply to show readers how ecocritical literary reading can be applied. If the reader is a teacher, this study can be a model for their students to practice. A student who reads this work is also expected to have a general idea of ecocriticism. If the reader is not a teacher or a student but an enthusiast, she or he is also expected to have an opinion on the awareness-raising function of ecocriticism. To make a brief introduction to ecocritical analysis, I examine *White Noise*. This section will help the reader apply ecocriticism to video games.

White Noise is the story of Jack Gladney, a college professor at the College-on-the-Hill in a small American town, who is the chair of Hitler studies and lives with his wife, Babette, and their children from previous marriages: Heinrich, Steffie, Denise, and Wilder. When discussing *White Noise*, it is worth mentioning the enormous difference between its beginning and ending. *White Noise*'s world of predisaster and postdisaster shows what impact a disaster is capable of inflicting. Today, we have seen what a medical disaster is capable of due to the COVID-19 epidemic. Though perhaps less acutely experienced, modern-life activities such as the use of detergents, the use of genetically engineered foods, and exposure to pollutants, irritants, and carcinogenic factors obviously have an effect on human health and the climate. As a consequence, pollution and waste are part of a multifaceted issue that has a detrimental effect on both human and nonhuman life. *White Noise* reflects a kind of technology which becomes dangerous. For Heidegger, a kind of technology which abstains from intervening in the natural flux of events is acceptable. For example, it is acceptable to build a windmill to use the wind power as the windmill operates in harmony with the natural flux of the wind. However, it is not acceptable to build a dam as it intervenes the natural flow of water.

> The revealing that rules in modern technology is a challeng[e] . . . which puts to nature the unreasonable demand that it supply energy that can be extracted and

stored as such. But does this not hold true for the old windmill as well? No. Its sails do indeed turn in the wind; they are left entirely to the wind's blowing. But the windmill does not unlock energy from the air currents in order to store it.[17]

Heidegger's statement asserts that technology is an element that makes life easier when it is environmentally friendly, but that it can be dangerous when it is done in a harmful way. Attempts to modify the natural systems of the ground, water, and soil ultimately result in destruction. Ivan Illich points out how technology in everyday life acts as a mechanism to mold people's thinking patterns and to teach them how to think about their needs.

> Welfare bureaucracies claim a professional, political, and financial monopoly over the social imagination, setting standards of what is valuable and what is feasible. This monopoly is at the root of the modernization of poverty. Every simple need to which an institutional answer is found permits the invention of a new class of poor and a new definition of poverty. Once basic needs have been translated by a society into demands for scientifically produced commodities, poverty is defined by standards which the technocrats can change at will. Poverty then refers to those who have fallen behind an advertised ideal of consumption in some important respect.[18]

Welfare and poverty definitions are measured as surpassing and falling below the requirements set by technocrats. A capitalist conception of technology brings in class distinctions and deepens the existing inequality against classes. That is to say, if we accept technological advancement as a facilitator of life, it will eventually fall behind social development for people who belong to a class that cannot afford it. Therefore, technology's efficacy should be questioned.

In *White Noise*, there is such a view of technology, especially in the pre-disaster period, that is, when the characters have not awakened to reality. Gladney, the main character of the novel, is a professor who thinks that thanks to technology and prosperity, he will not suffer natural disasters. Talking about natural disasters, Gladney claims that "these things happen to poor people who live in exposed areas. Society is set up in such a way that it is the poor and uneducated who suffer the main impact of natural and man-made disasters."[19] Here, Gladney takes a correct point of view, dividing disasters into natural and man-made. But it is wrong to think that such disasters will only affect the poor. Gladney ironically claims that he will not be affected by disasters, asserting that "people in low-lying areas get the floods, people in shanties get the hurricanes and tornadoes. I am a college professor. Did you ever see a college professor rowing a boat down his own street in one of those TV floods."[20] In fact, it is disturbing for Gladney, an intellectual, to have such a point of view. The opposite is expected: he must resist this injustice. But his way of speaking seems to be content with this injustice and to enjoy feeling privileged. Educators should discuss at this moment with students what the author is attempting to achieve, as well as what he could be attempting to help them realize in their own lives. Teachers can directly ask the question, "Is there a difference between the rich and the poor in terms of exposure to natural disasters? Is there a difference between the rich and the poor? Is the fact that they live in more secure homes enough to protect the rich from natural disasters?" These questions can open a meaningful classroom discussion.

Jack Gladney's ordinary life changes as a result of an extraordinary event. A disaster called the Airborne Toxic Event, a catastrophe caused by a gas released into the air as a result of a train accident, changes the lives of the characters in the novel. Gladney and his son Heinrich comment on the issue after hearing about the accident. While discussing the event, they imply that human and nonhuman beings will be affected differently from this event:

> "The radio calls it a feathery plume," he said. "But it's not a plume."
> "What is it?"
> "Like a shapeless growing thing. A dark black breathing thing of smoke."
> "Why do they call it a plume?"
> "Air time is valuable. They can't go into long tortured descriptions."
> "Have they said what kind of chemical it is?"
> "It's called Nyodene Derivative or Nyodene D. It was in a movie we saw in school on toxic wastes. These videotaped rats."
> "What does it cause?"
> "The movie wasn't sure what it does to humans. Mainly it was rats growing urgent lumps."[21]

The exchange reflects a general human-centered perspective. So why is this technology admiration so evident in the work? Does Don DeLillo criticize this? Should the awakening of the characters after the event be read as a catharsis? These are some of the many questions an ecocritical analysis tends to ask. It is a good idea to write about the background of the work, regardless of whether the writer purposefully or coincidentally created the setting. It is the ecocritic's duty to unearth the implications of the text. Therefore, I provide here a simple historical context on *White Noise*.

THE BACKGROUND OF *WHITE NOISE*

The year 1981 was a turning point in American political history. The republican candidate Ronald Reagan was elected president having taken 51 percent of the votes. The actions Reagan took included a reduction of taxes, an increase in the military budget and a deterioration of social security and the health system. "In the summer of 1981, Congress passed the Economic Recovery Tax Act, the largest tax reduction in U.S. history."[22] With this act, the amount of taxes that poor people had to pay decreased from 14 percent to 11 percent. At first glance, this act could be seen as a positive change; however, the amount of taxes that wealthy Americans had to pay reduced also, which made it difficult for the American economy to sustain itself. Another reform was enacted in 1986 that enabled people to pay even less than the above-mentioned amounts. With this new act, the amount of taxes that wealthy businesspeople had to pay dropped dramatically to 35 percent. Reagan arguably worked in favor of industry and against the working class. Although the economy was revitalized in general, the poverty rate rose dramatically. Therefore, the Reagan years were beneficial for the rich and a disaster for the poor. American policies in general ran parallel to this. A system which works for industry normally favors the upper class and ignores the needs of the lower class. Reagan defended capitalism, free trade, and individualism. He once made a speech claiming that "trees cause more pollution than automobiles do."[23] Reagan also reduced the budget of Environmental Policy Affairs and dismissed three EPA experts from service. Frank O'Donnell, director of Clean Air Trust states that "the administration tried to cut EPA funding by more than 25 percent in its first budget proposal."[24] C. Brant Short declares in his book *Ronald Reagan and the Public Lands: America's Conservation Debate, 1979–1984* (1989) that after Reagan was elected president he supported "the Sagebrush Rebellion" which enabled the industry to use public lands.[25] This act meant that public lands would be sold to industry. With reduced EPA budgets, much less importance attached to environmental problems, and a roaring industry favored by the government, environmental disasters were inevitable.

The fact that Ronald Reagan won a victory in 1981 elections in the United States, Helmut Kohl was elected Germany's chancellor, and the decline of communism in various countries including Czechoslovakia, Romania, and Poland with the Warsaw Pact meant that world politics was turning rapidly to neoliberalism. Neoliberal policies urged that technology should be improved; as a result, technological competition among countries turned into technological war. With a constant urge to be superior to other countries, new technological developments were advanced with greed. It is probable that necessary precautions were not taken to eliminate the unexpected consequences of technology since there was no time for this in the heat of the race. The novel critiques a flawed understanding of technology and an unconditional faith in technology. It puts forth the irony of humanity by stating that people "could put [their] faith in technology. It got you here, it can get you out. This is the whole point of technology. It creates an appetite for immortality on the one hand. It threatens universal extinction on the other. Technology is lust removed from nature."[26] It is possible to infer that the novel criticizes the perception of technology of the time it was written.

A TECHNOLOGICAL DISASTER: AIRBORNE TOXIC EVENT

Lives are turned upside down when a train car derails, triggering the novel's central issue: an airborne toxic event. The novel is divided into two main story lines: the airborne toxic event causing the release of toxic gases and the trial of the new psycho-pharmaceutical, Dylar. When Jack Gladney decides to use the unlicensed drug Dylar to deal with his post-traumatic fear of death, the two perspectives become entangled in the book.

It is possible to assert that DeLillo has a critical attitude toward the perception that some human beings have a privileged position in the world, understood from the fact that Gladney underestimated natural catastrophes and thinks that they only affect poor people and regions. In the end, Gladney understands that the mechanism of disasters does not differentiate between the rich and the poor. In fact, humanity is only one part of the environment, and if the environment is affected, the human body is also affected.

After having simply introduced the story, a teacher or lecturer can offer some of Glotfelty's research questions for the class to carry out a discussion. For instance, Glotfelty's first research question is an ideal one to ask high school students, since the question does not require expertise:

How Is Nature Represented in This (Work)?

Although the discussion session is open to any kind of ideas as long as they are relevant to the question, I personally believe that *White Noise* represents nature as neutral. Throughout the novel, there is an abundance of toxicity. The reader encounters, for instance, an ironical setting with "people trekking across wasted landscapes"[27] where they get exposed to toxicity. Toxicity is one of the most significant themes in the novel.

There are so many substances and wastes mentioned that nature cannot be expected to tolerate such a toxic burden. While all this is happening, it is interesting to discuss with students the main character's claims that natural disasters will only harm the poor. Refuting this claim, the fact that there was a disaster in his own community and Gladney's life was shaken to the ground can be interpreted as DeLillo's criticism of this anthropocentric perspective. All these events trigger problems for Gladney such as a chronic fear of death, hypochondria, depression, and forgetfulness. To alleviate these concerns, he uses an unlicensed drug, adding an addiction to his troubles.

Gladney's experience of self-realization is one of the most important themes of the novel. In the person of Gladney, this general point of view is criticized. Although Gladney is an educated person who experiences good living standards, he is equally affected by the disaster. Gladney claimed the contrary and mocked the idea of disasters hitting the rich; however, his experiences make him accept the vulnerability of human beings to disasters regardless of their social class.

At this point I suggest referring to Jane Bennett and other important contemporary theorists. Let us take Bennett's statements as an example. Bennett states that the interrelated relationship between human and nonhuman beings is at its highest point so far (2010), noting that "there was never a time when human agency was anything other than an interfolding network of humanity and nonhumanity; today this mingling has become harder to ignore."[28] For this reason, we must abandon the delusion that human beings are superior to other living things.

As El mentions in *Slow Violence in Contemporary American Environmental Literature*, "the way science and humanities examine the environment . . . has dramatically changed in the last three decades" and "human-centered approach[es]" are being "abandoned."[29] It is time to keep up with this change. Since young people are the ones who should reflect this change and awareness in every aspect of life, hoping that this study will attract their attention, the next section is devoted to video games.

INSTRUCTIONAL RATIONALE

Reset Earth is a video game launched by the United Nations. As the name suggests, it aims to raise awareness among young people of the possible consequences of climate change. In order to remind users how close the disasters that climate change will bring are if urgent action is not taken, and to raise awareness on this issue, the game establishes the year 2084 as its setting. A short, animated movie was also produced by the United Nations Environment Program. As part of the project, the United Nations released a trailer with the game. In the trailer, the character named Knox introduces herself by saying "This is me, Knox. Just another average girl living in a post-apocalyptic waste land."[30] There is also a very important warning for the future. As we see in the COVID-19 outbreak today, the outbreak of a new disease forms the premise of the video game. This production, which aims to raise the awareness of young people, although it may seem a bit hyperbolic, states that this disease reduces human lifespan to the age of twenties. While the prediction may seem exaggerated, it highlights an important contrast. Normally, our hopes for the future include that human life will increase thanks to the development of technology—a common anthropocentric point of view. This cautionary prediction, just like the disaster that befalls Gladney in *White Noise*, reminds us of our delusion that such incidents will not happen. In fact, although such disasters may seem remote, this game has retracted apocalyptic time and kept human lifespan exaggeratedly short to remind us of the consequences of inaction. The movie shows the grave of a person who lived only eighteen years between 2049 and 2067 in the video.[31]

People die young, suffering from a deadly skin disease called "the grow," which is caused by the thinning of the atmosphere. Knox and Sagan travel back in time to 2055 to investigate the foundations of this disease and find a solution.[32] The image of the world that Knox encounters in 2055 is a world where technology reaches its peak. Flying cars, which can be considered symbols of high technology, are seen everywhere.

In this film, young people's desire to return to the past and save people should be evaluated as a message given to today's young players. Although it is not possible to go back to 2055 from 2084 and save the world from disasters, it is possible in the year 2021 to strive to prevent the disasters that may occur in 2055. Knox and her team determine that the disease manifested by skin cancer and blindness in humans is caused by radiation from depletion of the ozone layer. Knox perceives the significance of this situation and says: "If we can stop these gases, we can stop the ozone layer from being destroyed."[33] After they learn that the hole in this ozone layer was caused by CFCs that emerged in the 1920s, they want to go back to that time and prevent the depletion of the ozone layer.

Fate of the World (Red Redemption Ltd.) is another video game which aims to raise awareness about climate change. In the opening trailer for the

Figure 4.1. The Grave of an Environmental Victim.
Source: UN Environment Programme, *Reset Earth*, 2020, screenshot, https://ozone.unep.org/reset-earth.

Figure 4.2. Knox's Travel in Time Back to 2055.
Source: UN Environment Programme, *Reset Earth*, 2020, screenshot, https://ozone.unep.org/reset-earth.

video game, it shows the year 2020 to refer to the enormous catastrophe experienced last year and states that the world had ignored climate change.[34]

The first mission of *Fate of the World* is to bring prosperity to Africa. It emphasizes how much Africa needs environmental justice, which the Environmental Protection Agency defines as "the fair treatment and meaningful involvement of all people regardless of race, color, national origin, or income, with respect to the development, implementation, and enforcement of environmental laws, regulations, and policies."[35] However, the situation in Africa is obviously an infringement of the rights of African people and the exploitation of land, both in real life and in the context of this game. Both race and income are significant factors of this infringement, and it is fair to

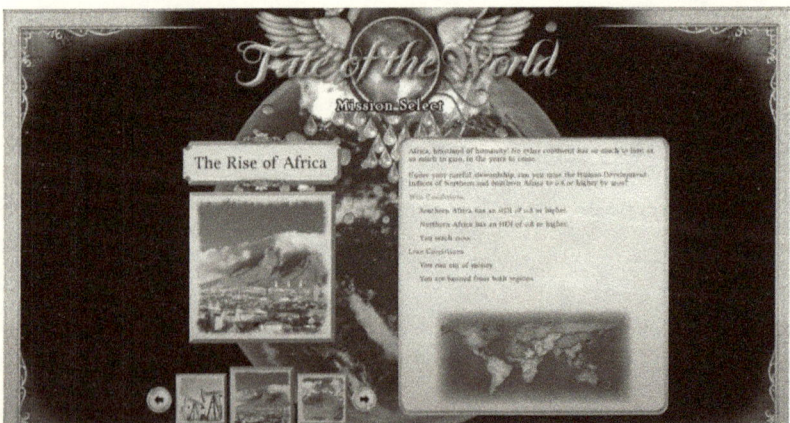

Figure 4.3. The Beginning of the Trailer of *Fate of the World*.
Source: Red Redemption, *Fate of the World*, 2010, screenshot, https://www.youtube.com/watch?v
=1AuIOGPJ9tw.

say that environmental racism is apparent in Africa. It is a moral imperative to call for sustainable development, prosperity and healthy living conditions in Africa; therefore, this game is also very useful from a pedagogical point of view because of the role empathic critical thinking plays. The second mission is called "Oil Fix It,"[36] where the player unearths oil to save money to use for the environment. As it may be expected from a game with environmental consciousness to promote wind and solar energy, the player might feel conflicted about this mission. Whether the game is critiquing this or trying to encourage the player's critical thinking is not very certain at first. However, the connection builds very logically allowing the players to understand the harm of fossil fuel extraction. Although unclear at first, the game critiques fossil fuel extraction and allows the players to understand for themselves what this process will cost. That is to say, nonenvironmentally friendly transactions harm nature regardless of whether they are made with good or bad intention. Therefore, humanity must calculate its impact on nature and the environment in every process. The third mission, called "Fuel Crisis,"[37] deals with this environmental disaster. At this point, humanity is on the brink of a crisis. Oil extraction reaches its peak. The mission is to reach the year 2120 before human extinction occurs due to global warming.

Importantly, this game has a part called "Denial."[38] This episode reminds us of what will happen if we ignore climate change. As climate change denialism is common, recognizing that this is wrong is one of the most important steps to take. Another mission of the game is called "Three Degrees,"[39] in which the player aims to keep the global heat below three degrees. Since global warming is the biggest environmental concern of our day, this section is also very significant since it both warns the player about the consequences of global warming and teaches them its causes. Another episode called "The Flood"[40] explores how deadly floods caused by human-made disasters can be, while "Dr. Apocalypse"[41] is aimed at keeping the environment at a normal temperature that can sustain human life.

By introducing these game scenarios into classrooms, students will acquire knowledge in the form of learning by doing. Students will not memorize the causes and effects of global warming; they will discover for themselves how global warming happens, what can be done to address it, and how they can help prevent further warming. The second benefit is that students will learn through these games because they are interested. Since playing video games is a common habit of many students, teachers and parents have the opportunity to help their students and children choose educational games and avoid harmful ones.

Considering the readiness level of the class, it is appropriate to apply this study between eighth and twelfth grades. The age range for the games is on average between twelve and sixteen. Although gaming and ecocriticism seem far from each other, a narrative lens makes it possible to examine any text. Just as an advertisement, or even a poster, can be examined from a literary point of view, it is extremely possible and worthwhile to examine video games from an ecocritical perspective.

The aim is to raise awareness of climate change. The use of video games for this purpose is very relevant for young people and offers the engaging hook of perspective-taking and virtual participation in problem-solving. The expectations of educators can be twofold: students can learn by enjoying video games and they also avoid violent contexts by playing such pedagogically suitable games. The students will be informed about the facts of climate change and its consequences as they learn to become environmentally-conscious citizens.

The Expected Results of Learning through Gaming

1. Students will acquire knowledge about climate change.
2. Students will be informed about the dynamics of global warming.
3. Students will be informed about the probable consequences of environmental disasters.
4. Students will gain habits, such as going to school on a bicycle, avoiding air conditioners, and consuming less energy, which help reduce impacts of climate change.
5. Students may participate in campaigns advocating for environmental rights.
6. Students may actively participate in social media events about climate change.
7. Students may start their own environmental campaigns.

Teachers are encouraged to support students' comparison of their level of consciousness before understanding the literary theory ecocriticism and before engaging with these games and the level of consciousness they experience after this new learning experience. Students are asked to present projects on global warming, detailing its reasons and what they think can be done to avoid further warming. The Fridays for Future movement should also be encouraged at school. Students are asked to find other climate change games and make a similar presentation about them. Students who successfully complete all of these tasks are given a badge with the words "eco-friendly student" on it.his lesson is supposed to raise awareness of climate change and make students more proactive about it. It is expected that after having done the tasks determined in this lesson, students will learn the causes of climate change and realize what to do and what to avoid in order to minimize its impacts. In this study, games are specially chosen because they attract student interest. The main field of work in ecocriticism is the written text, but it is not limited this. Animated movies, cartoons, documents, movies, and video games can be the subject of research for ecocriticism. Most importantly, after students have acquired this ecocritical perspective, they will be encouraged to view other texts in this way.

NOTES

1. Rachel Carson and Linda Lear, "Introduction," in *Silent Spring* (Boston: Houghton Mifflin, 2002), xvii.

2. James Lancelot Andrewes and John Field, "Genesis," in *The Holy Bible* (Cambridge: Printed by John Field, Printer to the University, 1668), chapter I.

3. Anonymous, *The Epic of Gilgamesh*, (Penguin Classics, 2000), 1.

4. Anonymous, *The Epic of Gilgamesh*, 2.

5. William Rueckert, quoted in Glotfelty, Cheryll, and Harold Fromm. *The Ecocriticism Reader: Landmarks in Literary Ecology.* (Athens: University of Georgia Press, 1996), 107.

6. Rueckert, *Landmarks in Literary Ecology*, 114.

7. Rueckert, *Landmarks in Literary Ecology*, 113.

8. Cheryll Glotfelty and Harold Fromm, *The Ecocriticism Reader: Landmarks in Literary Ecology* (Athens: University of Georgia Press, 1996), xviii.

9. Glotfelty and Fromm, *The Ecocriticism Reader*, xix.

10. Glotfelty and Fromm, *The Ecocriticism Reader*, xix.

11. Camilo Gomides, "Putting a New Definition of Ecocriticism to the Test: The Case of the Burning Season, a Film (Mal)Adaptation." *Interdisciplinary Studies in Literature and Environment* 13, no. 1 (2006): 16.

12. Richard Kerridge and Neil Sammells, *Writing the Environment: Ecocriticism and Literature.* (London; New York: Zed Books, 1998), 5.

13. Simon Estok, "Shakespeare and Ecocriticism: An Analysis of 'Home' and 'Power' in King Lear." *AUMLA* 103 (May 2005): 16–17.

14. Scott Slovic, "Forum on Literatures of the Environment." *PMLA* 114.5(1999): 1102.

15. Slovic, "Forum on Literatures of the Environment," 1102.

16. Slovic, "Forum on Literatures of the Environment," 1102.

17. Martin Heidegger,. *The Question Concerning Technology and Other Essays.* Translated by W. Lovitt (New York: Garland, 1977), 14.

18. Ivan Illich, *Deschooling Society* (London, New York: Marion Boyars, 2002, ©1970.), 3.

19. Don DeLillo, *White Noise.* (London: Picador, 2011), 114.

20. DeLillo, *White Noise*, 114.

21. DeLillo, *White Noise*, 109.

22. Michael Schaller, *Ronald Reagan* (New York: Oxford University Press, 2011), 44.

23. "Do Trees Pollute the Atmosphere?," *The Guardian* (Guardian News and Media, May 13, 2004), https://www.theguardian.com/science/2004/may/13/thisweeksscriencequestions3.

24. O'Donnel quoted in Romm, "Reagan Helped Save the Ozone Layer but Ruined America's Leadership in Clean Energy," *Grist*, April 2, 2021, https://grist.org/article/2011-02-07-reagan-helped-save-the-ozone-layer-but-ruined-americas/.

25. Calvin Brant Short, *Ronald Reagan and the Public Lands: Americas Conservation Debate, 1979–1984.* (College Station, TX: Texas A&M University Press, 1989).

26. DeLillo, *White Noise*, 285.

27. DeLillo, *White Noise*, 122.

28. Jane Bennett, *Vibrant Matter a Political Ecology of Things.* (Durham: Duke University Press, 2010), 31.

29. El, Erden. *Slow Violence in Contemporary American Environmental Literature.* (Newcastle upon Tyne: Cambridge Scholars Publishing., 2020), 31.

30. "Reset Earth: One Ozone. One Planet. One Chance," YouTube, accessed March 3, 2022, https://www.youtube.com/watch?v=WP8h3IahGCE.

31. YouTube, "Reset Earth."

32. YouTube, "Reset Earth."

33. YouTube, "Reset Earth."

34. "Fate of the World on Steam," Fate of the World on Steam, accessed March 3, 2022, https://store.steampowered.com/app/80200/Fate_of_the_World/.

35. "Learn about Environmental Justice," EPA (Environmental Protection Agency), accessed March 3, 2022, https://www.epa.gov/environmentaljustice/learn-about-environmental-justice#:~:text=Environmental%20justice%20(EJ)%20is%20the,environmental%20laws%2C%20regulations%20and%20policies.

36. "Fate of the World on Steam," accessed March 3, 2022, https://store.steampowered.com/app/80200/Fate_of_the_World/.

37. "Fate of the World on Steam."

38. "Fate of the World on Steam."

39. "Fate of the World on Steam."

40. "Fate of the World on Steam."

41. "Fate of the World on Steam."

BIBLIOGRAPHY

Anonymous. *The Epic of Gilgamesh*. Penguin Classics, 2000.

Bennett, Jane. *Vibrant Matter a Political Ecology of Things*. Durham: Duke University Press, 2010.

Carson, Rachel, *Silent Spring*. Boston: Houghton Mifflin, 2002.

DeLillo, Don. *White Noise*. London: Picador, 2011.

"Do Trees Pollute the Atmosphere?" *The Guardian*. Guardian News and Media, May 13, 2004. https://www.theguardian.com/science/2004/may/13/thisweeksscriencequestions3.

El, Erden. *Slow Violence in Contemporary American Environmental Literature.* Newcastle upon Tyne: Cambridge Scholars Publishing, 2020.

Estok, Simon. "Shakespeare and Ecocriticism: An Analysis of 'Home' and 'Power' in King Lear." *AUMLA* 103 (May 2005): 15–41.

"Fate of the World on Steam." Fate of the World on Steam. Accessed March 3, 2022. https://store.steampowered.com/app/80200/Fate_of_the_World/.

Glotfelty, Cheryll, and Harold Fromm. *The Ecocriticism Reader: Landmarks in Literary Ecology.* Athens: University of Georgia Press, 1996.

Gomides, Camilo, "Putting a New Definition of Ecocriticism to the Test: The Case of the Burning Season, a Film (Mal)Adaptation," *Interdisciplinary Studies in Literature and Environment* 13, no. 1 (2006): 13–23. doi:10.1093/isle/13.1.13

Heidegger, Martin. *The Question Concerning Technology and Other Essays.* Translated by W. Lovitt. New York: Garland, 1977.

Illich, Ivan. *Deschooling Society.* London, New York: Marion Boyars, 2002, ©1970.

James, Lancelot Andrewes, and John Field. "Genesis," in *The Holy Bible.* Cambridge: Printed by John Field, Printer to the University, 1668.

Kerridge, Richard, and Neil Sammells. *Writing the Environment: Ecocriticism and Literature.* London; New York: Zed Books; New York, 1998.

Lear, Linda, and Rachel Carson. "Introduction," in *Silent Spring.* Boston: Houghton Mifflin, 2002.

"Learn About Environmental Justice | Environmental Justice | US EPA," n.d. Accessed March 3, 2022. https://www.epa.gov/environmentaljustice/learn-about -environmental-justice.

"Reset Earth: One Ozone. One Planet. One Chance." YouTube. Accessed March 3, 2022. https://www.youtube.com/watch?v=WP8h3IahGCE.

Roark, James L. et al. *The American Promise: Volume B, A History of the United States.* Boston: Bedford/St. Martins, 2012.

Romm, Joseph. "Reagan Helped Save the Ozone Layer but Ruined America's Leadership in Clean Energy," *Grist*, April 2, 2021, https://grist.org/article/2011-02 -07-reagan-helped-save-the-ozone-layer-but-ruined-americas/.

Schaller, Michael. *Ronald Reagan.* New York: Oxford University Press, 2011.

Schram, Martin. "Nation's Longest Campaign Comes to an End," *Washington Post*, November 4, 1980.

Short, Calvin Brant. *Ronald Reagan and the Public Lands: Americas Conservation Debate, 1979–1984.* College Station, TX: Texas A&M University Press, 1989.

Slovic, Scott. "Forum on Literatures of the Environment," *PMLA* 114.5 (1999): 1102.

Chapter Five

A City for the Future

*Designing Socially Just,
Sustainable Urban Environments
with Elementary Students*

Alexandra Laing

The first day of school is an invigorating rush—full of adrenaline, antici-
pation, and nerves. Bundled within the upcoming academic year are new
students and parents, standards to be met, discipline to be managed, lunch-
time tutorials to be had, and laughter to be heard. Yet beyond that is the
weightiness of the need to prepare the students sitting in classrooms for a
climate-challenged world. In just the last few decades, climate change has
become a national issue. It began with Dr. James Hansen, the then director
of NASA's Institute for Space Studies in Manhattan, providing testimony
in 1988 before the United States Senate Energy and Natural Resources
Committee where he brought forth clear, data-backed evidence and stated,
"I would like to stress that there is a need for improving these global climate
models, and there is a need for global observations if we're going to obtain a
full understanding of these phenomena."[1] The *New York Times* ran an article
the very next day with the headline, "Global Warming Has Begun."[2] It is not
inconsequential that more than thirty years later, scientists are still referenc-
ing Dr. Hansen's testimony.[3] As an educator, with the catalysts to innovation
sitting in the desks of the classrooms we are entrusted, empowering students
with the knowledge to confront climate crises rests in the learning experiences we are able to provide. Yet, these experiences are not often richly
provided in our curriculum or nested in the standards that guide standardized
testing. This presents a challenging but not impossible obstacle for educators.

This chapter explores a standards-based, year-long inquiry of the envisioning and prototyping of new, innovative cities that run on sustainable energy to support citizens in the face of global climate change that also ensure social justice and support anti-bias government. This learning experience has students engaging in collaborative teams through four distinct phases to develop a new city (see figure 5.1). This inquiry differs from many open-inquiry experiences in the sense that it is framed in a foundation of integrated content standards through an inquiry-driven, problem-based learning[4] experience that utilizes technology as a platform for learning. Through the city's development phases, students investigate the use of tidal barrages to harness hydroelectric energy, study traffic congestion and the use of alternate public transportation methods, develop a proposal to combat coastline erosion, and make a plan to support a healthy economy in their new city through entrepreneurship and strategic city planning by combating the declining bee population and its impact on sustainable agriculture.

The City of Ember[5] by Jeanne DuPrau is a realistic, futuristic, fictional take on what the world would be like without the knowledge of natural light. The novel follows two young teenagers as they explore their relevance in society and question the structures that have been put in place by the founders of the city. With this learning module being highly interdisciplinary and lessons learned in one subject area supporting the learning in other subject areas, this chapter is laid out by phases. Within each phase, an overview of the purpose is first addressed to provide an in-depth understanding of the learning goals and the project. Then a standards overview is presented. Finally, each subject area is described through the lens of the exploration timeline of the main content which leads into the discussion, evaluation, and closing. For consistency, English language arts is presented first in every phase, followed by science, mathematics, and social science and history, respectively. While the inquiry is designed to be approachable through isolated subject areas, the power of the overall study is found in the interdisciplinary approach through the application of cross-disciplinary study. The standards that are used as the framework

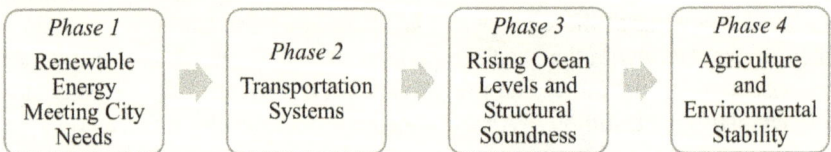

Phase 1	Phase 2	Phase 3	Phase 4
Renewable Energy Meeting City Needs	Transportation Systems	Rising Ocean Levels and Structural Soundness	Agriculture and Environmental Stability

Figure 5.1. A City for the Future, Yearlong Inquiry Overview. This yearlong inquiry is designed to be used with students in fourth or fifth grade. The engineering, science, and mathematics concepts can be scaffolded up or down according to student readiness or to fit other grade levels. If adapting for other grades, the literary lens may need to be altered accordingly to support students' instructional reading levels.
Source: Alexandra Laing

for this year-long inquiry are state-specific but are referenced in terms of general learning targets and included in detail in the lesson plan resource to represent the range of skills that can be addressed through this problem-based learning approach. Standards can and should be adapted to meet the program objectives for the school site and location as needed.

Throughout the four phases, there are specific, targeted lessons and action steps to engage students in climate change advocacy. Targeted lessons include: exploring renewable energy sources and tidal barrages, understanding key shifts in transportation infrastructure including maglev[6] train innovation and traffic congestion's impact on the climate and global citizens, real threats of weathering and erosion globally due to climate change, understanding climate-conscious decision-making and the implications of choice, the current global problem of the declining bee population,[7] and the current global issues surrounding food production.[8] The inclusion of social justice standards[9] have also been woven throughout this year-long inquiry. Targeted lessons include: analyzing the doctrine of discovery,[10] addressing Eurocentrism, studying the effects of the mindsets promoted by discoverers, city zoning, redlining, building highways through residential areas and historically black neighborhoods, and vegetation deserts.

Educators should think through each of the four phases in the dynamic way in which one phase builds on the next. However, if one phase or one lesson does not fit with the overall intent of the standards being addressed in your classroom, that's okay! The four phases and the individual lessons have been written in a way in which each can exist in solidarity, but the strength of this unit lies in the intentional and careful scaffolded development of topics. The greatest learning for students will occur with a true implementation of this year-long study.

PHASE 1: RENEWABLE ENERGY MEETING CITY NEEDS

In phase 1, students will investigate how renewable energy can meet a city's needs. *The City of Ember*[11] investigates how the city is powered by the moving water of the river because of the generator. In this phase, students have the opportunity to investigate a newer type of generator technology called a tidal barrage that uses the ocean's rising and falling tides to generate electricity. Students will design a new city and figure out a way to bring power to the city with hydroelectric energy. Students will form a team of designers and city planners to create this new city design that will be built along a water source, either relevant to a community location or identified as an ideal development area by national or global standards. A successful city plan will include a city layout blueprint that fulfills all aesthetic requirements identified, an explanation of how hydroelectric energy is being harnessed to power the city, a 3D printed or other model of a tidal barrage to be used along the city's water source, and a movie to "sell" the development of the city to the city contractors and builders.

In phase 1, which is all about renewable energy meeting the city needs, students investigate the topics of key ideas and details, writing opinion texts and conducting research in English language arts. Students investigate the nature of science and renewable energy in the sciences. They investigate lines, planes angles, symmetry, and two-dimensional figures and polygons in mathematics. Finally, students investigate the use of primary and secondary sources as it relates to explorers to America and geographic features in the social sciences and history. Technology components used include research strategies, 3D printing, modeling blueprints with digital technology, and the integration of visual presentation skills.

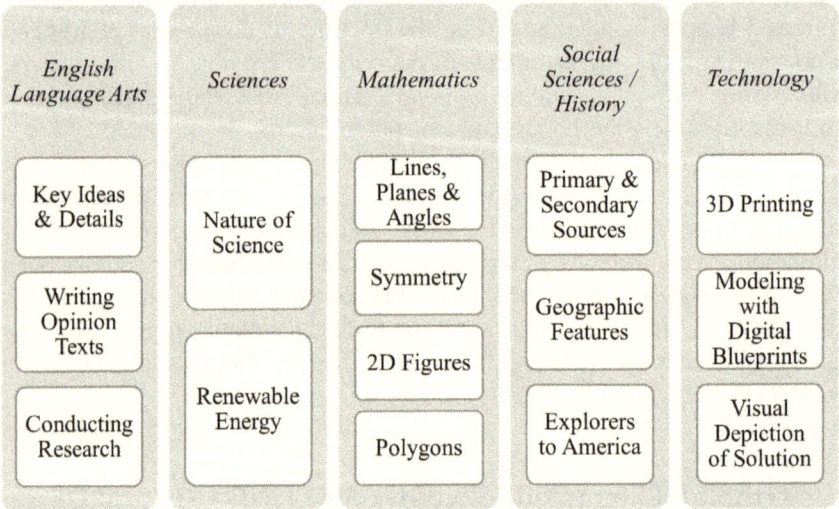

English Language Arts	Sciences	Mathematics	Social Sciences / History	Technology
Key Ideas & Details	Nature of Science	Lines, Planes & Angles	Primary & Secondary Sources	3D Printing
Writing Opinion Texts		Symmetry	Geographic Features	Modeling with Digital Blueprints
Conducting Research	Renewable Energy	2D Figures	Explorers to America	Visual Depiction of Solution
		Polygons		

Figure 5.2. Interdisciplinary Content Focus for Phase 1: Renewable Energy Meeting City Needs.

Source: Alexandra Laing

English language arts has students working to infer and synthesize to develop theories about primary and secondary characters and investigate how their actions impact events in fiction and nonfiction literature through the lens of The *City of Ember*.[12] The check for understanding with English language arts is to have students write an outline for an opinion essay, using a boxes and bullets method to provide the reader a rationale of why life in their city is better. The mentor text, *The City of Ember*,[13] provides deliberate word choice and models that can be used to engage students with thought-provoking components for the reader. For further study, students can develop their boxes and bullets outline into an essay, utilize peer review, or even publish their essays on a digital blog, website, or around the school campus to showcase their work.

In the sciences, students explore and compare and contrast renewable and nonrenewable energy sources to understand the distinction between them and identify the benefits and drawbacks of renewable sources. In the sciences, the check for understanding is a culminating activity where students investigate and explain what tides are, describe how those tides are related to the revolution of the moon around the earth, and describe the tides' role in a tidal barrage. Students will also design and print a 3D model of a tidal barrage and explain how a tidal barrage can use the movement of the tides to generate energy that can be converted into usable electricity. This activity also pulls on the information from the renewable and nonrenewable resources study, and students will be able to explain how a tidal barrage can use the movement of the tides to generate energy and convert that into usable electricity. Students can document the 3D printing process of their model title barrage, with a time lapse recording for further study.

In mathematics, students use an image of a local ecosystem to trace right angles, acute angles, and obtuse angles. Students also identify angles by modeling, creating, and labeling a shape with prescribed angle values. Students investigate parallel and perpendicular lines to identify how lines do and don't ever cross each other and to create and label a shape using angles and lines. Finally, students understand symmetry by folding a piece of paper in half, drawing a random shape, and opening the folded paper to reveal the line of symmetry through the middle. Students use hands-on investigation of polygons to describe polygons according to their symmetrical components. In mathematics, the check for understanding brings students to explore the characteristics of two-dimensional figures through the development of a city blueprint by blending mathematics and cartography literacies. Students will design and explain their city by imagining an aerial view from a drone or airplane. In designing the city blueprint, students will meet inclusion expectations, using angles, parallel and perpendicular lines, and polygons.

In the social sciences and history, students analyze the doctrine of discovery[14] that motivated explorers during the exploration and discovery of the New World (which was old to the indigenous inhabitants), colonial times, or westward expansion. Students will be able to address Eurocentrism by explaining the significant effects of the mindsets that were promoted by "discoverers" which became problematic in Western culture and led to genocide. Educators can check for understanding by having students compare and contrast the impacts that exploration and discovery posed for communities, geographic regions, and animals and plant life. It is important to have students investigate the impacts of the doctrine of discovery,[15] the ethical implications of action, and the responsibility of all citizens to speak up against injustice.

PHASE 2: TRANSPORTATION SYSTEMS

In phase 2, students will identify how studies of traffic congestion have raised concerns about an important public issue that impacts the daily lives of citizens. Students get an opportunity to design a city, in the same way that the City of Ember[16] was designed by its founders. Unlike the City of Ember, this new city will be designed with forward-thinking technology. In this phase, students will study problems caused by traffic congestion that are both practical and environmental and will design an alternate transportation plan for their city. Students will be challenged to figure out a way to reduce traffic congestion in their newly designed city with an innovative transportation system. The student team of designers and city planners will identify planned design for an updated transportation system for their city that uses magnetism and will build a small-scale working model. A successful design will include an updated city layout blueprint that illustrates the integration of the new transportation system, a working small-scale model of the proposed transportation system, a data-supported explanation of an improvement to the city's safety, and a commercial to inform and educate the citizens of the city about the new transportation system.

In phase 2, students use textual evidence to interpret and explain information, conduct research, and write informative texts in English language arts. In the Sciences, students are investigating the nature of science and properties of matter. In mathematics, students use patterns to solve multistep problems and investigate multiplication by one-digit numbers. In the social sciences and history, while the standards are state specific, they highlight the important connection of identifying relevance of a study to the location and area of the school environment. The standards listed in the lesson plan are an example of how a local study can be integrated into phase 2. Alternately, a variety of locations can be chosen for this study, based on the connection to the literature and climate awareness. Students are also encouraged to study local government as well as technological advances that impact their state or region of study. Technology integration includes robotics and coding, modeling with digital blueprints, and the use of a visual depiction of the solution.

In phase 2, in English language arts, students use researching skills to explore the history of transportation and identify key shifts in the automotive industry to compare the history of the transportation to *The City of Ember*'s[17] transportation structures. The check for understanding for English language arts has students explain how city safety is improved by well-functioning

English Language Arts	Sciences	Mathematics	Social Sciences / History	Technology
Using Textual Evidence	Nature of Science	Patterns	Local Government	Robotics and Coding
Interpret / Explain Information				Modeling with Digital Blueprints
Writing Informative Text		Solving Multi-Step Problems		
	Properties of Matter		Advances in Technology	
Conducting Research		Multiplying by 1-Digit Numbers		Visual Depiction of Solution

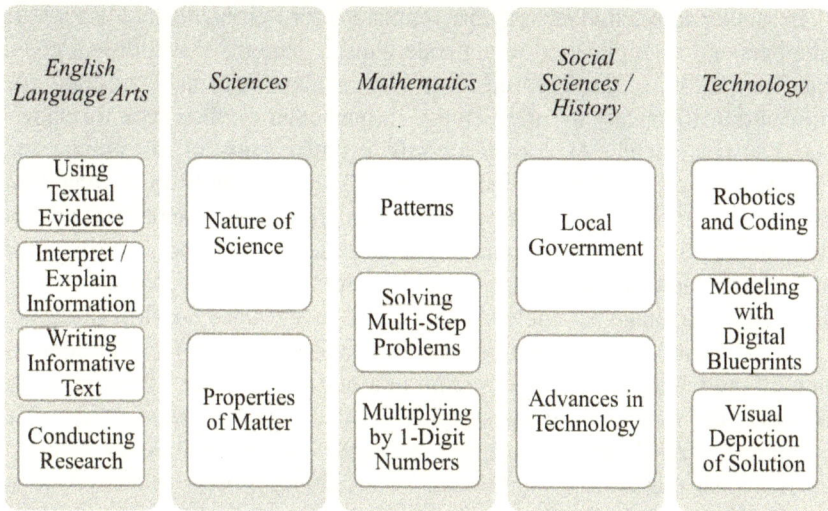

Figure 5.3. Interdisciplinary Content Focus for Phase 2: Transportation Systems.
Source: Alexandra Laing

transportation and has students identify how their newly developed transportation system will benefit the efficiency of their city. Students will use *The City of Ember*[18] as a mentor text to support their reasoning with examples from the text. Students will write an informational essay explaining how their new transportation system improves their city safety and efficiency by reducing congestion. For further studies, students can publish their essays on the digital blog, website, or around the school campus as a way to showcase their work. Students can also invite parents, community members, or other students to a showcase to demonstrate the history of transportation.

In the sciences, students explore and observe the characteristics of the properties of matter through classification stations and open investigations to measure volume and mass. Students also have the opportunity to explore the physical property of magnetism and describe how magnets are used in unique ways within industry. The check for understanding leads to the culminating project where students form into teams of designers and city planners to design a transportation system for their city that uses magnetism, similar to maglev[19] trains in the way they float above the track. Students will build a small-scale working model and explain the improvement the new transportation system brings to the city through the way it reduces the city's carbon footprint. Students should also explain the benefits of public transportation by using data to support their explanation. Further study can have students create a movie and commercial to inform and educate the citizens of the city about the new transportation system.

In mathematics, students explore multistep word problems that use whole numbers and multiplication to generate number patterns that follow a given rule of multiplication. This ties into how students can analyze commute time and traffic safety in intersections. In mathematics, the check for understanding has students exploring patterns in traffic statistics to estimate and calculate the congestion that will occur in their city. Based on the analyzed data and the application of their mathematics principles, students will look at the comparison between commute time and intersections and compare local and national statistics. For further study, students can utilize an online traffic simulation to change parameters and manipulate and study the flow of traffic.

In the social sciences and history, students will explore traffic congestion by modeling a traffic intersection in a busy spot on the school campus. Based on the simulation, students will work to design improvement in traffic patterns as a result of their learning. Students will utilize direction cards to instruct their motion and then record the modeling process to analyze it as a class afterward. In the social sciences and history, students will check for understanding by exploring different safety rules that are linked to city transportation and describe how some of these rules can be improved using robotics or coding to create a model and optionally program a device to improve city safety. Based on research about traffic congestion, students' physical modeling of a four-way intersection, and discoveries about their city safety design, students will design and build a model of a device that will improve city safety. This could be a new device, or an improvement to a currently existing city safety device. Students can engage in further study through the integration of LEGO WeDo 2.0[20] technology to build a replica by using the design library-based models and coding to ensure that their concept for a safer city can be clearly modeled.

PHASE 3: RISING OCEAN LEVELS AND STRUCTURAL SOUNDNESS

In phase 3, students will study how rising ocean levels are causing weathering and erosion to occur along the banks of the town's river which is putting the structural soundness of the town hall building in danger. Students will analyze the real and imagined threats to the citizens in *The City of Ember*[21] and will draw a comparison to the real threat of climate change impacting their own climate-conscious city. In this phase, students will understand that some threats are irreversible. Students will form into engineering teams to design a way to stop the weathering and erosion from further endangering town hall. A successful solution will include an updated city layout blueprint that shows the location of town hall and applies conversion of measurement to illustrate the distance town hall is located up-river, a digital or physical prototype or description of the proposed solution, and a presentation to explain the solution proposal, its strengths and weaknesses, and the reason the solution will be effective.

In phase 3, students investigate key ideas and details and the craft and structure of text in English language arts. In the sciences, students investigate the nature of science, weathering and erosion, and rocks and minerals as it relates to changing climate effects. In mathematics, students focus on multiplying multidigit numbers, as well as converting measures and solving for distance in a real-world problem-solving scenario. In the social sciences and history, students study the geographical features of their state to understand climate change. With the application of technology, students have the opportunity to extend their thinking through modeling with digital blueprints and the utilization of visual depictions of solutions.

In phase 3, English language arts students compare the real and imagined threats in *The City of Ember*[22] to the real threats of climate change and coastal erosion. Students will use researching skills to explore occurrences of weathering and erosion around the world with a variety of student-friendly articles including "Living on Edge: Californians Watch as Cliffs Crumble Way toward Homes"[23] and "Kiribati: The Face of Climate Change"[24] from Newsela[25] and will identify methods that industries have used to combat these problems.

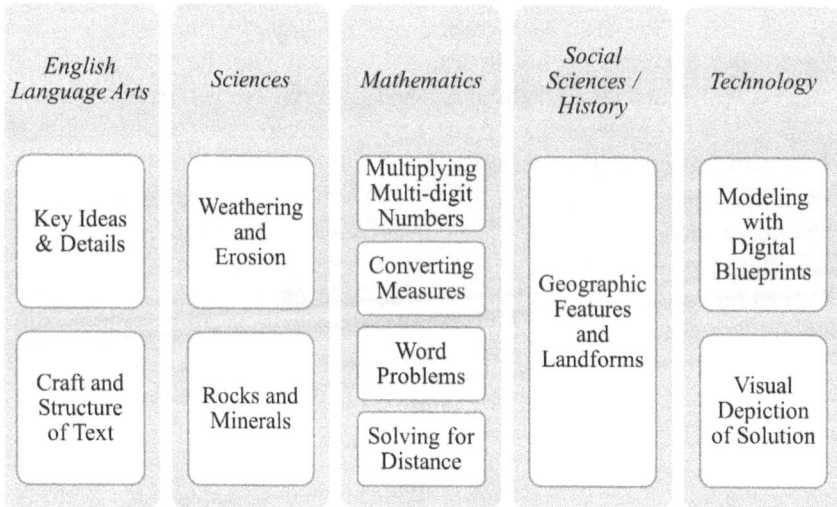

English Language Arts	Sciences	Mathematics	Social Sciences / History	Technology
Key Ideas & Details	Weathering and Erosion	Multiplying Multi-digit Numbers		Modeling with Digital Blueprints
		Converting Measures	Geographic Features and Landforms	
Craft and Structure of Text	Rocks and Minerals	Word Problems		Visual Depiction of Solution
		Solving for Distance		

Figure 5.4. Interdisciplinary Content Focus for Phase 3: Rising Ocean Levels and Structural Soundness.

Source: Alexandra Laing

The check for understanding in English language arts in phase 3 has students exploring how engineers are trying to creatively solve weathering and erosion problems around the world. Students write an informational essay about what would happen to the world if weathering and erosion, as caused by climate change, disappeared. Students should plan and organize their writing, create a rough draft, and peer-conference with others to identify revisions and create a final draft. Students can further their study by publishing their essay on digital blog website or around the school campus in a showcase of their work.

Within the sciences, students investigate the natural processes that cause weathering and erosion and simulate these processes through several different models including chemical weathering, mechanical weathering, and erosion. Students will create a virtual gallery of examples of chemical and mechanical erosion models that are found around the world. To check for understanding, teachers can present students with images and ask students to identify correct examples of chemical weathering, mechanical weathering, and erosion. Critically, students should not only be able to correctly identify examples but also describe how weathering and erosion are related. This learning can be enhanced with integration of the arts by having students act out the forces that cause weathering and erosion in a visual arts performance.

Within mathematics, students explore conversion of measurement by applying a scale to their city layout and determine the rate of change that impacts their self-designed town hall building. Students are provided with a scaled erosion rate and asked to calculate the actual erosion rate using conversion. This illustrates a real-world problem many cities around the world face through the crisis of climate change by applying mathematic principles. Students will demonstrate their understanding through a performance task in which they will create their own scale using conversion of measurement to mathematically depict their understanding. Students can further their study by creating a diorama to showcase their to-scale model, explaining how understanding mathematical size and scale benefits understanding.

Within social sciences and history, students will utilize research to describe the need for and the characteristics of renewable and nonrenewable resources that are found near or around their own geographic location. Students will extend their thinking by exploring technologies that are being developed to support the use of these renewable resources and predict how these technologies could change the local or global economy. Students will check for understanding by explaining how climate-consciousness benefits this generation and generations to come. Students will deepen their understanding by investigating the implications of choice, making connections to climate-conscious decision-making.

PHASE 4: AGRICULTURE AND
ENVIRONMENTAL STABILITY

In phase 4, students will study the current worldwide problem of a declining bee population.[26] They will analyze the impact this real-world problem will have on their city's food production and economic stability and will understand the need for the Sun. The City of Ember[27] is a city with no sun, and the characters in the story are unable to tell time without a clock. The gathering hole clock measured the hours of night and day. It was never allowed to run down, and the job of the timekeeper was to wind the clock every week and place the date sign in the square every day. In this phase, students draw comparisons to the way bees function in a hive to the way that characters in the story interact and depend on each other. Students will also investigate the way the sun informs a bee's travel and will design a sundial for their schoolyard. The telling of time is a critical component of everyday life for all creatures. Helping students develop context for understanding the passing of time closely connects with a students' ability to understand how change unfolds. With the sundial, students understand the interaction between the Earth's rotation and the Sun's energy. Students will be challenged to update the previously designed city garden and identify how to combat the declining bee population through the garden, increase food production, and create a small business by selling the honey from local hives to boost the city's economy. A successful plan will include an updated city layout blueprint that shows the location of the new business' headquarters and the integration of hives into the garden and around the city; a webpage or other advertising platform for the new business that uses effective advertising, identifies the product and its related cost, and describes the business' contribution to solving the worldwide problem of the declining bee population[28]; a demonstration of how a bee moves, communicates, and pollinates flowers using robotics; a commercial advertising the new honey brand; and a sundial installation created through an art collaboration in the school garden.

In phase 4, students study English language arts with the integration of knowledge and ideas. In the sciences, students focus on plant living systems while also studying pollinators and their role in food production. In mathematics, students apply concepts of angles to real-world scenarios within nature. Within the social sciences and history, students study economics to understand how businesses can earn a profit and the risks and rewards provided to entrepreneurs. Students extend their thinking through technology integration with website creation, robotics and coding, modeling with digital blueprints, and creating visual depictions of their solutions.

English Language Arts	*Sciences*	*Mathematics*	*Social Sciences / History*	*Technology*
Integration of Knowledge and Ideas	Living Systems - Plants / Pollinators and their Role in Food Production	Angles	Economics and Business	Website Creation / Robotics and Coding / Modeling with Digital Blueprints / Visual Depiction of Solution

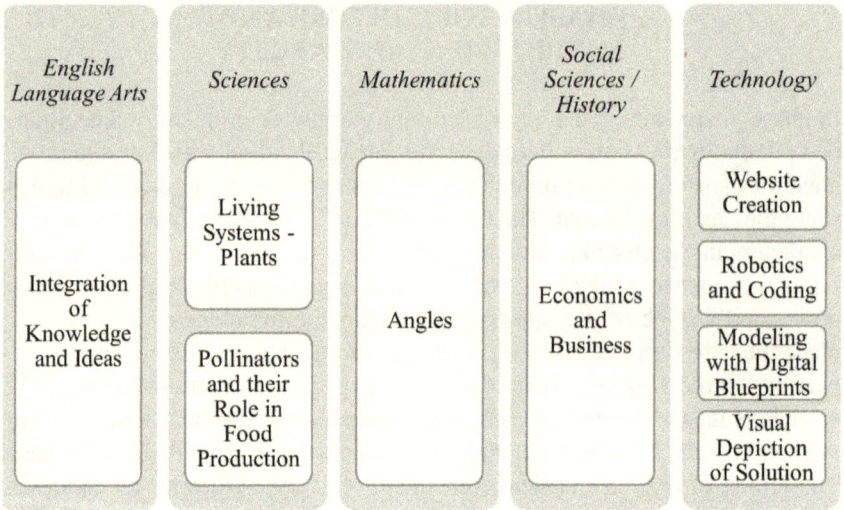

Figure 5.5. Interdisciplinary Content Focus for Phase 4: Agricultural and Environmental Stability.
Source: Alexandra Laing

Within phase 4, students study English language arts through comparing and contrasting the organization and interdependence of bees in a hive to the way that the characters in *The City of Ember*[29] operate. This is a unique approach to understanding character development by studying the natural phenomena of the organization of a bee colony to identify similarities and differences to characters in *The City of Ember*.[30] The check for understanding for English language arts in phase 4 is to have students write an informational essay about what would happen to the world if the bees disappeared. This activity connects all subject areas together and engages students in higher-order thinking and analysis. Students should follow successful writing norms of planning and organizing their writing, creating a rough draft, and conferencing with others to identify revisions and create a final draft.

In the sciences, students explore and observe how bees work through coding with the Dash robot[31] technology and the Blockly[32] coding app to collect pollen and make honey which is a hereditary behavior that also simultaneously pollinates plants. Students investigate the important physical structure of bees and look to understand how bees and their behaviors positively impact the environment. Students identify the connection between bee behavior and pollination and leads them to engage with the real-world problem of the declining bee population. Students discover the critical role bees play in global food production around the world which is impacted drastically by climate change and the loss of habitat. In the sciences, a check for understanding has students investigating how different plants invite specific pollinators to visit them and demonstrate how pollen is transferred to other flowers. Students will model pollination with several methods and choose a flower for a specific type of pollinator to develop a model with available materials. Classes should conduct gallery walks so that students can observe characteristics of multiple kinds of flowers, noting the common function of flower parts and the differences in pollination needs.

In mathematics, students explore and recognize the measurement of angles by not only studying the composition of beehives to investigate how angles can be additive, but by practicing measuring angles with images and real-world items. Students can extend their thinking using robotics and the Blockly[33] coding app to program a Dash[34] robot and illustrate angles. Educators can have students check for understanding by hosting a shadow study throughout the course of a day to observe a shadow's movement in relation to the sun's through the sky. Students will enrich their understanding by studying how sundials track the arc of the sun as it moves along its apparent path as seen from a fixed location on Earth.

In the social sciences and history, students will describe how entrepreneurs can start a new business and identify the factors that help new businesses become successful. Through the development of a new small business, students will identify how to combat the declining bee population[35] with a city garden to increase food production and sell honey from local hives to boost the city's economy. Students will extend their thinking and have the opportunity to engage creatively with technology by creating a website for their small business that uses effective advertising, identifies their product and its related cost, and describes their business's contribution to solving the worldwide problem of the declining bee population. In the social sciences in history, students should check for understanding by investigating the current social and global issues of food production in decline by engaging in research with articles such as "Spray Meant for Zika Mosquitos Kills Millions of Bees,"[36] "Creating a Buzz—Teens Find Sweet Success with Black Bee Honey,"[37] "Robotic Bees Could Spread Pollen to Plants If Real Bees Die,[38]" "Honeybee Food Hot Spot Is No Longer Hot,[39]" and "Insects, Animals That Pollinate Plants Are Dying Off, Report Says"[40] from Newsela[41] and the video documentary, "Queen of the Sun."[42] Additionally, students will investigate how cities are often heat zones due to the lack of shade from trees. With an extension to climate and environmental justice, students investigate the tie of sparse vegetation to low-income and typically marginalized communities where students will engage in discussion of heat and light sharing.

Table 5.1. Evaluation Rubric

SCORE	Content Knowledge	Engineering Design	Technology Integration	Originality, Value, and Style of Products
4	I have designed: • a way to apply the content in new and unique ways • a controlled experiment to test new theories	I explored each problem and designed a best solution that: • considered all constraints • was a new, unique solution • considered additional factors not identified by the problem	I choose my own technology to: • create product(s) • gather new information • engage my audience	The product(s) I created: • include all required components • were a unique and high-quality product • show a personal touch • would work in the real world
3	I have demonstrated proficiency by: • applying accurate principles from science, math, ELA, and/or social studies	I explored each problem and designed a best solution that: • considered all constraints • was a well-developed solution	I use technology as asked to: • create product(s) • gather new information • engage my audience	The product(s) I created: • include all required components • are well-made and designed with style
2	I have started to demonstrate an understanding of some of the content, but I have some misconceptions.	I explored each problem and designed a solution that: • considered some of the constraints • was a solution that could work but that needed additional revision	I use some of the technology to: • partially create product(s) • gather new information • engage my audience	The product(s) I created: • include most of the required components • have some interesting design
1	I have not yet demonstrated an understanding of most or all of the content, and I need more support.	I started to explore each problem and thought about how to develop a solution but did not succeed.	I do not apply the technology to: • create product(s) • gather new information • engage my audience	The product(s) I created: • are missing a lot of required components • are unfinished or unoriginal

CONCLUSION

Empowering students with the understanding of their local and global impact on climate change and social justice is an often-absent topic from classrooms. With this engaging and thoughtful interdisciplinary approach to these topics, educators are able to address their content area standards and the important topics that are impacting our world today. Nesting the year-long study in a comparative analysis of literature simultaneously promotes literacy and real-world engagement. The hands-on nature of these lessons enhances student engagement which leads to deeper understanding. Highly rigorous checks for understanding, performance tasks, and year-long projects that build upon each other demonstrate to students the interconnectedness of their learning to the real world. In a world that faces many challenges, engaging students in A City for the Future is doing more than teaching reading and math—it's teaching students to be global citizens who are climate-conscious, socially just, and anti-biased.

NOTES

1. "June 23, 1988 Senate Hearing 1," Pulitzer Center. https://pulitzercenter.org/sites/default/files/june_23_1988_senate_hearing_1.pdf.

2. Philip Shabecoff, "Global Warming Has Begun, Expert Tells Senate," *New York Times*, June 24, 1988. https://www.nytimes.com/1988/06/24/us/global-warming-has-begun-expert-tells-senate.html.

3. Peter Sinclair, "Judgment on Hansen's '88 Climate Testimony: 'He Was Right,'" *Yale Climate Connections*, June 20, 2018. https://yaleclimateconnections.org/2018/06/judgment-on-hansens-88-climate-testimony-he-was-right/.

4. "What Is PBL?" PBLWorks. https://www.pblworks.org/what-is-pbl.

5. Jeanne DuPrau, *The City of Ember*. New York: Yearling, an imprint of Random House Children's Books, 2016.

6. "How Maglev Works." Energy.gov. https://www.energy.gov/articles/how-maglev-works.

7. "US Beekeepers Continue to Report High Colony Loss Rates, No Clear Progression toward Improvement." Office of Communications and Marketing. https://ocm.auburn.edu/newsroom/news_articles/2021/06/241121-honey-bee-annual-loss-survey-results.php#:~:text=Beekeepers%20across%20the%20United%20States,Bee%20Informed%20Partnership%2C%20or%20BIP.

8. "Food." United Nations. https://www.un.org/en/global-issues/food.

9. "Social Justice Standards." Learning for Justice. https://www.learningforjustice.org/frameworks/social-justice-standards.

10. "Doctrine of Discovery." Upstander Project. https://upstanderproject.org/learn/guides-and-resources/first-light/doctrine-of-discovery.

11. DuPrau, *The City of Ember.*

12. DuPrau, *The City of Ember.*

13. DuPrau, *The City of Ember.*

14. "Doctrine of Discovery."

15. "Doctrine of Discovery."

16. DuPrau, *The City of Ember.*

17. DuPrau, *The City of Ember.*

18. DuPrau, *The City of Ember.*

19. "How Maglev Works." Energy.gov. https://www.energy.gov/articles/how -maglev-works.

20. "WEDO 2.0 Support: Everything You Need." LEGO® Education. https: //education.lego.com/en-us/product-resources/wedo-2/teacher-resources/teacher -guides.

21. DuPrau, *The City of Ember.*

22. DuPrau, *The City of Ember.*

23. "Living on Edge: Californians Watch as Cliffs Crumble Way toward Homes," *Newsela.* By Associated Press, adapted by *Newsela* staff, February 24, 2016. https:// newsela.com/read/house-cliff/id/15047/.

24. Maddie Rhodan, "Kiribati: The Face of Climate Change." *Newsela.* iGeneration Youth, adapted by *Newsela* staff, October 30, 2018. https://newsela.com/read/ kiribati-climate-change/id/46787/.

25. "What's New," *Newsela.* https://newsela.com/.

26. "US Beekeepers Continue to Report High Colony Loss Rates, No Clear Progression toward Improvement," Office of Communications and Marketing. https: //ocm.auburn.edu/newsroom/news_articles/2021/06/241121-honey-bee-annual-loss -survey-results.php#:~:text=Beekeepers%20across%20the%20United%20States,Bee %20Informed%20Partnership%2C%20or%20BIP.

27. DuPrau, *The City of Ember.*

28. "US Beekeepers Continue to Report High Colony Loss Rates."

29. DuPrau, *The City of Ember.*

30. DuPrau, *The City of Ember.*

31. "Wonder Workshop: Home of Dash, Cue, and Dot—Award-Winning Robots That Help Kids Learn to Code," Wonder Workshop—US. https://www.makewonder .com/robots/dash/.

32. "Home of Cue, Dash and Dot, Robots That Help Kids Learn to Code," Wonder Workshop. https://home.makewonder.com/apps/blockly.

33. "Home of Cue, Dash and Dot, Robots That Help Kids Learn to Code."

34. "Wonder Workshop: Home of Dash, Cue, and Dot."

35. "US Beekeepers Continue to Report High Colony Loss Rates."

36. Alan Yuhas, "Millions of Dead Bees Are a Side Effect of Spray Meant to Control Zika Spread," *Newsela. The Guardian*, adapted by *Newsela* staff, September 8, 2016. https://newsela.com/read/anti-zika-bee-death/id/21365/.

37. "Teen Entrepreneurs' Thriving Honey Business Creating a Buzz in Florida," *Newsela. Orlando Sentinel*, adapted by *Newsela* staff, April 3, 2018. https://newsela .com/read/teens-honey-entrepreneurs/id/41264/.

38. "Robotic Bees Could Be the Solution to the World's Bee Problem," *Newsela*. *The Guardian*, adapted by *Newsela* staff, October 16, 2018. https://newsela.com/read /robotic-bees-can-pollinate/id/46713/.

39. "Honeybees Finding It Harder to Eat at America's Bee Hot Spot," *Newsela*. Associated Press, adapted by *Newsela* staff, July 23, 2018. https://newsela.com/read/ elem-honeybee-trouble-great-plains/id/44763/.

40. "Want to Eat? Then Save Pollinators from Extinction, U.N. Report Says," *Newsela*. Associated Press, adapted by *Newsela* staff, March 14, 2016. https://newsela .com/read/pollinator-decline/id/15378/.

41. "What's New." *Newsela*. https://newsela.com/.

42. *Queen of the Sun. Queen of the Sun: What Are the Bees Telling Us*. A Collective Eye Production, 2014. http://www.queenofthesun.com/.

BIBLIOGRAPHY

Associated Press, adapted by *Newsela* staff. "Honeybees Finding It Harder to Eat at America's Bee Hot Spot." *Newsela*, July 23, 2018. https://newsela.com/read/elem -honeybee-trouble-great-plains/id/44763/.

————. "Living on Edge: Californians Watch as Cliffs Crumble Way toward Homes." *Newsela*, February 24, 2016. https://newsela.com/read/house-cliff/id/15047/.

————. "Want to Eat? Then Save Pollinators from Extinction, U.N. Report Says." *Newsela*, March 14, 2016. https://newsela.com/read/pollinator-decline/id/15378/.

Auburn University. "US Beekeepers Continue to Report High Colony Loss Rates, No Clear Progression toward Improvement." *Office of Communications and Marketing*. https://ocm.auburn.edu/newsroom/news_articles/2021/06/241121 -honey-bee-annual-loss-survey-results.php#:~:text=Beekeepers%20across%20the %20United%20States,Bee%20Informed%20Partnership%2C%20or%20BIP.

DuPrau, Jeanne. *The City of Ember*. New York: Yearling, an imprint of Random House Children's Books, 2016.

Learning for Justice. "Social Justice Standards." *Learning for Justice*. https://www .learningforjustice.org/frameworks/social-justice-standards.

LEGO® Education. "WEDO 2.0 Support: Everything You Need." *LEGO® Education*. https://education.lego.com/en-us/product-resources/wedo-2/teacher-resources/ teacher-guides.

Newsela. "What's New." *Newsela*. https://newsela.com/.

Orlando Sentinel, adapted by *Newsela* staff. "Teen Entrepreneurs' Thriving Honey Business Creating a Buzz in Florida." *Newsela*, April 3, 2018. https://newsela.com /read/teens-honey-entrepreneurs/id/41264/.

PBL Works. "What Is PBL?" *PBLWorks*. https://www.pblworks.org/what-is-pbl.

Pulitzer Center. "June 23, 1988 Senate Hearing 1," *Pulitzer Center*. https://pulitzer- center.org/sites/default/files/june_23_1988_senate_hearing_1.pdf.

Queen of the Sun. *Queen of the Sun: What Are the Bees Telling Us*. A Collective Eye Production, 2014. http://www.queenofthesun.com/.

Rhodan, Maddie. "Kiribati: The Face of Climate Change." *Newsela*: iGeneration Youth, adapted by *Newsela* staff, October 30, 2018. https://newsela.com/read/kiribati-climate-change/id/46787/.

Sinclair, Peter. "Judgment on Hansen's '88 Climate Testimony: 'He Was Right.'" *Yale Climate Connections*, June 20, 2018. https://yaleclimateconnections.org/2018/06/judgment-on-hansens-88-climate-testimony-he-was-right/.

Shabecoff, Philip. "Global Warming Has Begun, Expert Tells Senate." *New York Times,* June 24, 1988. https://www.nytimes.com/1988/06/24/us/global-warming-has-begun-expert-tells-senate.html.

The Guardian, adapted by *Newsela* staff. "Robotic Bees Could Be the Solution to the World's Bee Problem." *Newsela,* October 16, 2018. https://newsela.com/read/robotic-bees-can-pollinate/id/46713/.

United Nations. "Food." *United Nations.* https://www.un.org/en/global-issues/food.

United States Department of Energy. "How Maglev Works." *Energy.gov.* https://www.energy.gov/articles/how-maglev-works.

Upstander Project. "Doctrine of Discovery." *Upstander Project.* https://upstanderproject.org/learn/guides-and-resources/first-light/doctrine-of-discovery.

Wonder Workshop. "Home of Cue, Dash and Dot, Robots That Help Kids Learn to Code." *Wonder Workshop.* https://home.makewonder.com/apps/blockly.

Wonder Workshop—US. "Wonder Workshop: Home of Dash, Cue, and Dot—Award-Winning Robots That Help Kids Learn to Code." *Wonder Workshop—US.* https://www.makewonder.com/robots/dash/.

Yuhas, Alan. "Millions of Dead Bees Are a Side Effect of Spray Meant to Control Zika Spread." *Newsela: The Guardian*, adapted by *Newsela* staff, September 8, 2016. https://newsela.com/read/anti-zika-bee-death/id/21365/.

Chapter Six

Fostering Environmentalism and Activism in Students

Plastic Pollution as a Starting Point

Karen Ball and Elke de Vries

For younger generations, environmentalism is a legacy that will be handed over to them without a clear path toward a brighter future. It is the youth and children of today that will be left cleaning up the mess older generations have left. They will be the ones facing the harsh realities that generations before them have refused to acknowledge: that humans are having a significant impact on the climate and the environment around them. Because of this we need to give students a voice with which to change the present and ensure a better future.[1] Through our role as teachers, we should be able to inspire and encourage discussion about the environment and how we can address problems now. It is known that humans are unreliable at best when it comes to processing long-term risk, and as a result our handling of climate issues has been significantly impacted, leaving behind a heritage of woe for younger people (like our students).[2] Through teaching, we need to do better in promoting activism in the classroom so the people whose futures are at stake can have a say in how they are handled. One recurring issue that has attracted the attention of youth around the world is the use of single-use plastics and the detrimental effects that they are having on the environment. Developing a long-term plan to deal with an issue (like plastic waste) that will be around for hundreds of years is exactly the kind of planning students need to understand more about to develop real change for their future and that of their families. Therefore, this chapter will focus on three main sections—the first is empowering students in the classroom to become environmental advocates, including motivation and teacher attitudes. The second is narrowing the broad topic of environmentalism down to a subarea of plastic as a starting point to engage students. The final section goes deeper into developing meaningful lessons for students and engaging activism through literature.

Although there are many options to choose from when it comes to environmental sustainability concerns, the impact of plastics is something that is very controversial at the present time, with many students and youth of today already campaigning to end single-use plastics in various countries around the world.

SECTION I—SUPPORTING STUDENTS
AND EMPOWERING EDUCATORS

Introducing Activism

As educators, it is essential to understand environmental activism and bring this to light in the education world. Students today are passionate about this topic, and it is our responsibility as educators to bring that passion and consciousness into the school so that students understand that they are not alone and do not have to be only students at school and environmentalists outside of school. Even minimal research on the topic of environmental activism will reveal that the majority of people highly invested in this are the youth of today. A study done in the Nordic countries shows that "after nearly three decades of working with young environmental change makers, we can honestly say that the way THIS generation of young people is stepping up to solve the twin crises of climate change and deteriorating ocean health is unprecedented," said Leesa Carter-Jones, president, and CEO of Captain Planet Foundation, in a statement.[3] The study also reported that "most respondents say that their biggest concerns are plastic in the ocean (91%)."[4] The examples from this research are a great tool to use in the classroom.

Teachers are fortunate because, through the use of social media, literacy, and the internet, it is easier than ever to find material that will inspire and motivate students into activism. We do not need to reinvent the wheel to use environmental activism as a focus for lessons. Identifying local policies on single-use plastics (or whichever issue feels most appropriate) in a school community is a good first step.

Another important step in introducing environmentalism in the classroom is listening to students and asking what interests them. What do they know? What do they want to learn more about? What steps do they want to take in activism? Empowering students is also about listening and supporting their passions and concerns for the future. There may only be a few students who begin the discussion, but it often turns into a surprising conversation with the students you might least expect to hear from giving great input with other classmates feeling the impact and wanting to learn more as well. When students understand a problem and feel agency to solve it, they become advocates for change who dare to take action. As a teacher, it is difficult to leave a lesson like that without feeling pride in today's youth and having faith in the future generations. Once this conversation part is complete, it is then about the variety of sources used to encourage environmentalism in the classroom.

Local vs. Global

Teachers are encouraged to combine many different media to offer students the broadest possible lens through which to see the multiple facets of environmental activism. This is important for approaching the issues surrounding plastic waste, or any other environmental concern, so that students are given the opportunity to understand the various contexts involved with both the problem and its potential solutions. Students need to see not only the problem that is right in front of them but also how it impacts global concerns. That is why it is important to introduce both the local as well as global perspectives of environmental sustainability and activism. Educators can introduce the problem or environmental issue, in this case, single-use plastics, and also the history and the background of it. This is not only imperative from an educational viewpoint but also to ensure that history does not continually repeat itself. By using examples of global issues, students can see how different cultures and countries are dealing with particular issues and then compare those approaches with their own. Bringing it back to the local, students can investigate what is actually being done in their community or county, and come up with ideas about how to advocate for change. Including both perspectives will help teachers ensure they are having their students complete a thorough investigation of a particular factor of environmentalism. It is a wonderful starting point for researching and comparing students' enthusiasm and desires to push further into the topic and investigate the political issues behind environmentalism.

Politics and Activism

To further encourage environmental activism that has an impact, students need opportunities to see that there is a link between productive work and the ability to change the thoughts and minds of people in power behind the politics of environmentalism. This generation has shown that they are more aware of the political decisions being made and the ramifications that they are having on the future. Interviews taken with young sustainable change makers "showed they are very concerned about the complete lack of action from politicians, who do no more than talk about climate change. They see things getting worse, but still no action. The survey respondents feel frustrated that it is only young people taking action. The combination of increased awareness and knowledge about global climate crisis with lack of political action exacerbates the growing concern and worry."[5] Also important is the future voting habits of today's youth. From the Nordic study, of "the survey respondents, 40% report that they vote for political parties that prioritize climate or that they are engaged in a political youth organisation, and 21% state they are politically engaged for the environment in some other way."[6] This momentum should not be discouraged but developed through active lessons in schools. Educators can support this process by teaching students how to write better, research smarter, and express themselves for a specific purpose. Hopefully, teachers show students the positive impacts young people can have to change their future, using real-world examples. Sharing past examples of how students have enforced changes in politics is one way to reassure young people that past efforts have not been wasted. For example, a teacher could highlight the article "7 Times in History When Students Turned to Activism."[7] Such articles or videos can encourage students to stand up for their beliefs and give them concrete examples of how others have done so. It is important for students to see successful initiatives but also to understand the processes these individuals or groups of students had to go through to make a beneficial change.

Thinking positively about student agency and impact is a great incentive for students to help fight for change that will protect their future. This also strengthens the link between classroom content and real-world applications. In our modern world, students are able to communicate easily and globally in almost no time at all. The extent to which language can have an impact on people around the world for any number of different purposes has perhaps never been clearer.

Taking the first step toward activism and environmentalism is the first step in the right direction for many students, who may not realize the power of their voices. We hope to not only educate students about environmental issues but also to give them the tools they need to express themselves well, research reliable sources related to selected issues and participate in informed discussions. Students should never be dismissed as "too young" or "too inexperienced" because then we are taking away their right to have a say in their futures. We encourage teachers to believe in students and their ideas.

Although some ideas may be too broad or would take an extended time to achieve, it is important to help students, with a positive intention, attempt to break ideas down into manageable steps. Too often, students lose hope in their ideas too quickly because of the teacher's, rather than their own, discouragement. Listening to students and giving them a voice is key to empowering them. Celebrating small successes sounds like something that should be innate in educators' approaches today, but unfortunately, too often it is not. It is extremely valuable to acknowledge when a student achieves something, whether that be an excellent essay written about climate change, a moving poem, or mail from local government officials responding to letters a student wrote. Students are expecting and deserve this from teachers, who are their guides and directors of learning.

Teachers Are Powerful Advocates

To conclude the first section of this chapter, it is vital to remember that teachers have the power to support students and bring high value opportunities into the classroom at a time when ecological sustainability is so important. Being a teacher who steps up to address students' feelings about what is happening outside of the classroom may have long-lasting impacts on how students perceive their own agency. By looking at what is valuable to students, educators can modify existing curricular objectives, especially in the humanities, by tailoring the topics covered in these lessons and units to address students' interests and concerns. Even simple changes to a typical unit plan could have a lifelong impact on students and quite possibly the environment around them. As an example of this, the next section explores information for educators that focuses on plastic pollution, with an in-depth look at single use plastics.

SECTION II—THE TOPIC OF PLASTIC

This chapter segment explores the ways that plastic became such an instrumental feature in our lives. This unit with complete lesson plans as well as differentiation activities will be included on the resource tab of the website for this book. Plastic pollution is the lens through which students will be encouraged to be more active in their own community and also serves as an example for them to test their language and communication abilities. The hope is that it supports educators' efforts to prepare and challenge their students to take real world action. The development of plastic, an overview of plastic pollution, and information surrounding the debate of single-use plastics will be included in this section. The purpose of this is to support educators planning a unit of instruction around environmental sustainability with an in-depth focus on plastic and single-use plastics. A similar approach could be adapted for any number of topics.

In order to develop a well-planned unit, this approach is focused on one area of environmental protection that students at most grade levels could appreciate and demonstrate activism on—eliminating plastic pollution. The information provided in this section supports unit goals and tasks by outlining the path from the creation of plastic as a substance to its role in the environmental concerns that we are currently confronting.

As teachers, we noticed that as the topic of environmental sustainability has been growing in the media students around the world were taking action to fight further pollution. Although plastic pollution is related to climate change more in terms of our carbon footprint, we find it a relevant starting point with a tangible problem for students to grasp and become empowered by in their ability to participate and make change. By bringing real-life issues into the classroom and our instructional plans, we were able to create a unit that students could relate to and value inside as well as outside of the classroom. Recognizing that many students are passionate about climate change and global sustainability, this unit suggests ways to teach language skills and critical thinking, while providing ideas for supporting this important cause in students' everyday lives. First, it is important to highlight certain aspects of plastic development, including economic concerns that teachers and students will need to understand while completing this unit plan.

The Unintended Consequences of Plastic

Plastic was introduced in 1907, with the invention of something called Bakelite, a type of synthetic resin. It was used for many items and grew tremendously in popularity. Plastics were being used everywhere, and "by the end of the 20th century . . . had been found to be persistent polluters of many environmental niches, from Mount Everest to the bottom of the sea."[8] Plastic was praised because of its longevity—ironic given that this is the same reason it is so harmful to the environment. Plastic can take hundreds of years to break down.[9] Therefore, it builds up in places other than at designated disposal areas, such as city dumps or recycling centers. According to Greenpeace, "over 90% of plastics are not recycled," which means they remain as waste indefinitely.[10]

Issues Surrounding Single-Use Plastics

Single-use plastics are causing great concern for the environment. Plastic bags and food containers, for example, are used for only a short time before being tossed out, but they will exist as trash for many years beyond that of their single use. This has given humans in the last few decades the unfortunate title of being part of a "throw-away culture" whose trash problem is not merely about the space it occupies but the myriad impacts the breakdown of plastics can have on the environment and the animals that unwittingly ingest particles, including us.[11] Since plastic pollution has become a headline issue in the world, action is being taken to address it.

Many businesses have transitioned from plastic versions of commodities to offer paper straws, reusable bags, and paper wrappers. Some countries, such as Canada, have taken steps toward banning single-use plastics: the United Nations says that "more than 60 nations have taken steps to reduce single-use plastics by imposing bans or taxes."[12] Although the process to curb reliance on some plastics has begun, it will still take many years of hard work and people doing their part to end plastic pollution. Besides the new restrictions, there are still enormous amounts of these single-use plastics in our ecosystem, which will require clean up. Events such as World Cleanup Day and National Cleanup Day are normally held in September each year. "Initiatives like these are great ways to take action on environmental issues. Beach cleanups are not the only thing you can do to end plastic pollution across the globe, though— there are other ways to get involved through Earth Day Network's Great Global Cleanup, and End Plastic Pollution campaigns."[13] Educators speaking to students about these days, even organizing similar events can help inspire students toward environmental activism.

Plastics and Marine Life

Besides restricting the production and use of single-use plastics, there has been a tremendous effort to bring to light the issues regarding plastic disposal and the effects it has on marine life. The following information outlines the problems caused by microplastics and single-use plastics in relation to how it damages the marine life. It will also address what people can do to reduce the amount of plastic that ends up in the world's water sources.

Most of us have seen the terrible images of marine life that have been tangled in a plastic drink bottle holder or some other form of plastic that has been dumped into the water. Young people have grown up seeing these images and are aware that this type of dumping can cause real problems in nature. It is our job as educators to bring this part of their lives into the classroom. A note of caution when teaching about this subject is that some students could be very sensitive to any images of animals in distress. It is important to warn all students if graphic images are to be shown, or possibly avoid this, as the goal is to promote activism through awareness rather than through sadness or guilt. There are also other issues caused by microplastics that are not as blatantly visible as some of the images circulating the internet and news.

Microplastics

According to world animal protection.org, microplastics are "are tiny pieces of plastic which come from larger plastics that have degraded over time."[14] The major issue with microplastics is that sea life ingests them, and it makes them sick, causes them to suffer, sometimes leading to their death. The statistics regarding the amount of microplastics in the world's oceans are shocking.[15] For example, an article on how pollution is affecting seals and other marine life states "there are around 50–75 trillion pieces of microplastics in the world's oceans."[16] As this number continuously expands, it will be our current students and future generations who will need to make conscious choices and work toward reducing these staggering statistics. Students may ask where these microplastics come from and be surprised to learn that the biggest contributor of microplastics are from items such as plastic bottles and plastic shopping bags. With an average of approximately 80 percent of microplastics coming from these sources, students may feel motivated to brainstorm solutions regarding these types of everyday items and even to eliminate them in their own practices. [17]

Single-Use Plastics and Marine Life

A powerful connection for helping students of any age understand the concrete impact of these plastics is how they affect the world's sea life. Unfortunately, exorbitant amounts of this disposable commodity make it into the world's water sources. The 100 Ocean Plastic Pollution Statistics & Facts section of the helpful resource "It's a Fish Thing," states that "of the plastic waste produced annually, approximately 50%, or 419,980,609,462 pounds, is some form of single-use plastic" and "Globally, there are approximately 8 million pieces of plastic that enter the ocean every single day."[18] This statistic is overwhelming even before considering how detrimental this must be to the world's sea life, but we have opportunities to confront it. The next section will delve into this further.

Confronting the Issue

While the context for this significant problem is far more complex than any individual's choices, it is an appropriate starting point to help young people understand that we can all make choices every day to help prevent more plastics from harming the world's sea life. The Seattle Aquarium has produced a great blog that shows ways to support this cause. They suggest six ways that people can reduce the number of plastics in our everyday choices that range from using reusable bottles, bags, and straws to being conscious of plastic packaging when purchasing items to supporting World Oceans Day, which is June 8 each year.[19] Considering that "there are at least 700 species of marine animals that could go extinct due to ocean plastic pollution," discussing opportunities to combat statistics like this is an important aspect of any lesson about environmental impacts.[20]

Accessible facts and figures are important for providing students with actual data that inform them of the urgent issues surrounding sea life. It will help support the unit and also show that there are many ways to make a difference and reduce the amount of plastic used. By using class time to discuss these issues and how they can affect the planet and the future generations, students begin to understand that small choices do make a difference. Education is a great way to support environmental activism from the classroom to the wider world. The next section of the chapter will look at using the information about the environment to create meaningful lessons.

SECTION III—DEVELOPING MEANINGFUL LESSONS FOR STUDENTS

Engaging with Literature through Activism

Many curricula around the world are focusing more on activism through different subject lenses to offer students a more holistic view of environmentalism. The combination of skills that come from courses of study such as the International Baccalaureate's Language and Literature or Individuals and Societies ensures that students not only have important information about the environment, but also key competencies needed to enact real change in the world around them. The effort to apply effective language proficiency to specific topics across subject areas can be quite challenging for students, particularly if they have not had any experience working toward such parameters in their English classes yet. English language arts courses are usually based on general concepts or topics that permit adaptability for students. Instead, lessons should endeavor to create scenarios that challenge students to play an active role in their community. Through understanding and promoting activism in regard to a problem that is not only affecting them but the world at large, students can live up to the responsibility of caring for the earth. Grappling with climate-related problems will be this generation's challenge and educators can help prepare them for it. An unfair burden thought it is, students can be encouraged to confront it by channeling the momentum from other community activist groups into proactive learning experiences. Teachers can support this by being more creative in their lesson planning and using relevant examples from timely issues to engage students' deeper connection to learning.

As teachers, we can use a wide range of different literature types and genres for this purpose: protest poetry, novels, and documentaries can be very effective resources for helping students delve deeper into any topic. Ecocriticism is a study under which a variety of different literary types and ideas about the environment and its relationship with the world are considered. Simply put, "ecocriticism is a broad way for literary and cultural scholars to investigate the global ecological crisis through the intersection of literature, culture, and the physical environment."[21] Many people refer to ecocriticism as something that is difficult to define; however, it is focused on addressing ecological ideas through works of literature. This does not only have to be through the study of literary works but also through the lens of works in other subjects in the humanities and the sciences. There are numerous examples of authors whose writing helps students understand the effect humans have on the environment—authors such as Sarah Crossan, whose book *Breath* looks at the things in our environment that we take for granted, such as oxygen, or Colin Dann's *The Animals of Farthing Wood*, which is a classic tale about British animals seeking refuge after their homes are threatened by human development and the trials they face moving to a new "home." These stories help to show students multiple perspectives on similar problems that they may experience in their own lives.[22] This type of literature offers a great way for students to understand the effects of environmental issues on a personal level, even if the experiences are out of reach in their normal lives. Authors are increasingly using ecocritical approaches in their stories to help readers better understand environmental issues (see more examples in the lesson plan). Short stories can also be an interesting format for students to explore environmental concerns. *Everything Change: An Anthology of Climate Fiction* is a collection of twelve stories that examine different environmental issues and their effects on the people around them. These creative literary mediums offer wonderfully diverse ways for students to access complex environmental problems.

Poetry is another interesting literary genre that helps students explore environmental issues and the emotions associated with them. Protest poetry and poetry in general offer creative ways for students to engage with environmental concerns. Naomi Shihab Nye's "World of the Future, We Thirsted" or Kelly Roper's "A Choking Sky" are poems that deal with everyday occurrences from an ecological perspective.[23,24] These poems are meant to engage the reader to think more deeply about everyday environmental concerns and how people can react to them. "A Choking Sky" is a rhyming poem that blames factories for the pollution they produce and questions if there is not a safer way. Appropriate for older students, "World of the Future, We Thirsted" uses rhetorical questioning to make the audience reflect on the ways they have been treating the environment. Protest poetry is an effective genre to explore with students as it mixes the creativeness of poetry with the meaningfulness of environmentalism.

Students today are bombarded by problems in the media about how the world is changing and how there is little done to help improve severe pollution and related environmental problems. As we make decisions for our students' generation without their input, we could instead be enabling them with the proper skills they need to fight for their future. We want students to feel empowered so that they are able to go out into the world and do better for themselves and the next generations. Being able to show students how they can help not only themselves, but future generations is a powerful tool that we as teachers can help our students develop. We are in the lucky position of being able to show our students how to help themselves and to work for their future as opposed to passively accepting decisions that are being made for them. Students need to develop skills that will allow them to research, investigate, and exercise their right to choose how the environment that their generation will inherit is handled. Skills such as persuasion, creative thinking, and problem solving are invaluable for students to learn and practice. Coupling environmental issues with these skills helps students see how techniques learned in their study of literature can be applied to real world concerns.

Real life examples are the best way to communicate with students and to encourage them to act. Although we want to encourage imagination, it is important to have tangible examples and goals for students so that they can realistically put themselves into productive situations that contribute to actual change. Relevant real-life examples are usually effective with students and can be motivation enough. Plastic pollution is an ever-growing global issue that can be seen around the world in almost every community. With this example, students can see the real-world application of activism for change. As students perceive the power of their voice and the impact their words can have on others, they begin to recognize the responsibility they have to fight for their future. Some people believe that students are too young to fully understand the concepts they are protesting for; however, in some cases, it may be the fact that they are unable to properly articulate themselves. This is where the development of students' language skills for specific purposes is paramount in guiding them toward better expression of their ideas.

As the topics of climate change and global pollution grow in coverage by the world media, there are leading student activists who are serving as good role models. It is important to see how students are taking up the challenge to change the world that they live in for the betterment of their future, rather than being passive bystanders, or waiting until it is too late. Young adults such as Greta Thunberg, Xiuhtezcatl Martinez, or Wangari Maathai show others that there are better ways to make people in power pay attention to the decisions that they are making for the generation of students currently in classrooms. Greta demonstrated the power and momentum that could be harnessed through the organization of students by demonstrating in front of her parliament building every school day for three weeks. The movement "Fridays for Future" came from Greta's stance on confronting climate change and has evolved into a global phenomenon that is student-led.[25] It directly harnesses the power of students and their abilities to lead real change. There are strong, nonviolent, active examples of students like Greta leading the charge around the world on a variety of movements. Sharing this work with students in our own classrooms can help demonstrate the link between the skills that are taught in class and the skills needed to lead real change in the world.

Students need to see the value in their education in order to develop a deeper understanding that extends beyond the classroom. The link between persuasive speech and social issues is more relevant than ever before. As Greta Thunberg stated: "Why study for a future, which may not be there?"[26] As teachers, we have a responsibility to ensure that our students are prepared for the future that awaits them; this includes arming them with the fundamental skills of standing up for what they believe in and confidently preparing compositions worthy of the causes for which they fight.

Unit Description—Unit Title: Plastic, Not So Fantastic

Combining the idea of the impact of single-use plastic and the concepts of teaching real-world scenarios, this unit of work offers a broad look into the reality of environmental issues around the world combined with the skills that students need in order to hone their abilities to a real purpose. This also provides an opportunity to broaden the development of students' writing and analysis abilities regarding any number of humanities-focused topics. We want to be able to provide students not only the content and information about environmental issues but also the expertise with which to enact change on these issues through activism.

The unit of instruction we propose could be offered over six to eight weeks or developed further depending on the level of analysis desired. The teacher is able to use their discretion as to length and depth of study. The following explanation has been divided into different tasks, culminating in a summative task that looks at developing and using the skills practiced throughout the unit. A more detailed version of the plan is provided on the resource tab of the website for this book. Please note that any sources referenced will be specifically detailed in the tasks section below as well as in the lesson plan materials and bibliography. Also provided in the accompanying lesson plan is a list of sites to help teachers further research or find resources as needed.

Rationale for Lesson Approach

Educators can help students see that pollution is a key issue in today's world that affects not only one country or one area but exists as a global problem. To engage student empathy about the problem, they need to be more aware about how their lives and future are affected by the pollution that has been caused by generations before them. A unit of this nature is also a good demonstration of the way that people try to convince others to take notice and help, modeling for students how they might take similar action regarding other issues that may concern them. The study of language and literature, especially in the context of how humans behave or ought to behave in a society, plays a huge role in motivating, exciting and inciting young people to take action.

The humanities provide students with the content and context with which to understand the problem and investigate further, while a language focus provides the lens through which students view the content and also provides the tools with which to address it. This unit focuses on the subject of plastic pollution and the language technique of persuasion applied for middle school–level students.

Topic: Persuasive Writing

Students will explore the use of persuasive language in a constructive way that prompts change in their community—their task will be to convince their school community to use less plastic by organizing and producing texts that apply persuasive language techniques.

A text we use to anchor this unit is the 2017 documentary *A Plastic Ocean* directed by Craig Leeson.[27] This documentary is a wonderful educational tool for illustrating the detrimental effects that plastic waste is having in the ocean. It is a full feature-length documentary in which the creator/narrator travels around the world to document and spread awareness about the effects of our lifestyle choices. This documentary uses a range of examples, scientific studies, and community information to show the wide-ranging issues and problems that are a result of plastic pollution. Before viewing it in the classroom, students and parents should be cautioned that there are scenes which some students may find distressing. This includes, but is not limited to, injured animals, dead animals, animals in distress due to plastic entanglement, dissections of animals' stomachs, and animals being treated poorly.

This set of lessons has been developed to not only help students widen their understanding of pollution in our world, but also to expand on their persuasive writing skills, which are often employed in activism literature to incite change. Students are challenged to explore a range of literary texts to better explore the topic. The summative assessment at the end of the unit focuses on the literary skill of persuasive writing to stimulate action in their local community. While these lessons primarily set out to enhance English language skills, multiple crossovers to other subjects and collaborations with other teachers and subjects are possible.

Included in the unit are some examples of protest poetry, which combines a range of different literary techniques to address the issue of climate change and environmental action from a very different perspective. Many students have little to no experience with the literary form of protest poetry, making it an interesting way to get students involved.

A Plastic Ocean was created by Craig Leeson, an award-winning journalist, filmmaker, and public speaker who advocates for better education and action to prevent pollution in the oceans and the general environment.[28] He and his team have also made an educational discussion guide to accompany the documentary and help educators better access powerful messages within its storyline. All of these resources can be downloaded for free from the documentary's homepage. Some of the activities include word-search and crosswords from the movie.

This unit plan can be extended or shortened to the teachers' discretion and to meet students' needs. There are several additional programs, sites, and organizations that follow similar topics and provide students with different perspectives and alternative solutions.[29] We have tried to include as many links as possible, additional activities, and suggestions to help tailor this lesson plan to the needs of individual classes. We hope that you enjoy exploring this lesson with your class, and we hope it can encourage more students to think about their future, the environment, and other species.

Inquiry

A first step to introducing this or similar units is to ask students why they should care about the environment? Students should think about the ways in which we use the environment, but also what it can mean to us and our future. Students should make the connection that we all use the environment and therefore need to look after it, especially if they would like to preserve it for future generations.

Protest Poetry

After students have had some time to think about the environment and what might be important to them, they are going read how others express their ideas about the environment. Poetry can be a great way for students to explore new and different subjects. It is a shorter and more easily accessible literary option that can act as an introduction to the topic. It can also show students that there are a variety of topics poetry can address. We discuss protest poetry and its role in environmental activism, recognizing that the first question students might ask is "what is protest poetry?" Most students know the words *poetry* and *protest*, but seeing them together might be something new to them. A good definition might include the following: To protest means to object or to complain. Protest poetry or protest literature refers to writing that addresses real sociopolitical issues that exist in the world or within an author's local community. As mentioned by critic Jay Parini in the introduction to *Poems for a Small Planet: Contemporary American Nature Poetry*, "Nature is no longer the rustic retreat of the Wordsworthian poet. . . . [it] is now a pressing political question, a question of survival."[30] Poetry can itself express an objection against these issues. For instance, the poem "A Choking Sky," by Kelly Roper, or "A Lullaby in Fracktown" by Lilace Mellin Guignard are good examples as they deal with the refusal of some companies to take responsibility for their actions in contributing to global pollution. Other fantastic poems about environmental issues include "Epithalamia" by Joan Kane, "Waterdevil" by Jamaal May, or "The Greenhouse Effect" by Carl Dennis, just to name a few. Protest poetry should contain a message or "take-away" that the reader can keep with them.

The following poetry excerpts are examples for what students can use as their first experience with protest poetry. Students can read the full version of these texts and then ask themselves the following inquiry questions and discuss within the class:

1. How is imagery used to engage the audience?
2. How does the author try to "connect" with the topic?
3. How important is word choice in these examples?

Consider the poetry excerpts that follow:[31]

A Choking Sky	*It Starts with a Wrapper*
Watching smoke stacks choke the sky	It starts with a wrapper
Always makes me want to cry.	Casually dropped in the street.
I just can't help but wonder why . . .	Someone else will pick it up . . .

The next step is to challenge students with writing their own pieces of protest poetry. In order to do this, they need to research an environmental issue that is important to them or that they feel they have a personal connection with. This can be something local, national, or global. Students should be able to gather information on how the issue affects people, including history about the issue and narrowing it down to an aspect on which they can focus their poetry.

Students are then encouraged to form their own pieces of protest poetry. Although students should not be limited by length or a word count, they should be focused on the goal of creating a meaningful message within the poem. Final pieces could be written or performed.

An extension of this effort could be to study a similar concern in the context of a short story. Realistically, any short story could be used for this task. The main focus is to have a creative fictional story which illustrates the development of a climate-related problem and a character's response to it. Short stories are a wonderful way for students to see the lives of other people and experience the emotions of a wide variety of characters while taking a relatively small amount of instructional time. Unlike novels, short stories are easy to digest and can be set as homework tasks or short lessons. This task can be used with any short story about the environment, however excellent options are available from the collection of works *Everything Change: An Anthology of Climate Fiction.* Specifically, the story "Aqua Alta," by Ashley Bevilacqua Anglin, is a tale set in what appears to be the not-too-distant future which follows the journey of a young girl as she is forced to evacuate what is left of Venice with her great grandfather.[32] It is a story of hardship and being involuntarily moved due to the effects of climate change. The writing style is easy and friendly, mixing Italian words into the English prose. Anglin is a very emotive writer and her descriptions bring to light the emotionally difficult situation of leaving everything one knows behind.

Students might focus on the following points as they read:

- What types of symbols does Anglin use in her story? How do they correspond with the theme of the story?
- What do you think Anglin's view of climate change is?
- How would you feel if your home was taken away due to climate change?
- Who would you blame if this happened to you?
- Who should be responsible for people who are displaced due to climate change?

A follow-up to the discussion of these questions could be to have students write their own perspective of the short story based on a simple prompt like this: Write a mini story (400–500 words) that describes you or a character leaving home due to climate change catastrophes. Describe the environmental issue that has led to this, and highlight whether you feel there is a way to prevent it. Try to develop the emotions of your characters to show the consequences that climate change can have on people. In the accompanying lesson plan, we offer multiple options for differentiation, such as creating a poster or collage depicting the emotions felt by the characters.

After studying poems or short stories that express concern about an environmental issue, students may appreciate the documentary as yet another form of protest through storytelling. It is important for students to understand that a documentary is a nonfiction film that focuses on one important topic or issue, with the aim of trying to raise awareness. Documentaries are becoming increasingly popular in today's media as they are able to follow the journey of one person or one concern. This documentary follows the environmental issue of excess plastic in the world's oceans. Studying it has two purposes: it should give students a better understanding about the documentary text type as well as the way that writing/filming can be persuasive in encouraging people to act on certain subjects.

Introduction to Persuasion in "A Plastic Ocean"

Not only is the documentary itself an emotional experience for the viewer, the trailer provides a confronting and quite powerful experience.[33] It shows the story of how one man wanted to make a documentary about whales but turned out to be confronted by a much larger matter. It highlights the major concerns and issues that will be raised throughout the documentary as well as the persuasive techniques to engage the audience. In the opening comments and quotes, there are uses of evidence (eight million tons of plastic are dumped into our oceans), repetition (of the word *plastic*), inclusive language (*us* and *we*) and cause and effect reasoning (animals trapped in man-made plastic, we put the plastic into the ocean, the animals suffer). These are only the verbal and written forms of persuasion; the trailer also uses visual forms of persuasion, as it highlights the main confronting issues with stark images of plastic pollution. The director uses strong images of pollution juxtaposed with beautiful scenes of the ocean to enhance the connection between the pollution of the oceans and our wastefulness. In combination, the trailer employs an effective use of "shock value" to provoke a deep emotional response from the audience. This not only entices the audience to watch the documentary, but also serves to enhance and further instill the main message of the documentary: we need to be more aware of how plastic is being used in our day-to-day lives. Viewing the trailer with students before watching the whole documentary slowly introduces students to these ideas and messages while nudging them to consider how persuasive techniques are used in media.

In addition to viewing the trailer, it may be useful to have students consider short reviews of the documentary as they too offer aspects of persuasion.

"It brings the issue of plastic pollution **to your home. It's an eye-opener for all."**— *Clear Water Initiative via Facebook*

"The **most impor-tant film of our time."**—*Sir David Attenborough*

"It is both **confronting and heartbreaking but it informative and eye-opening."**—*The Ethical Straw Co. via Instagram*

Quotes taken from Hardin, Tod. "Resources." *Plastic Oceans International*, September 15, 2021, https://plasticoceans.org/plastic-pollution-info-resources/.

Students should look at how review quotes emphasize the importance of the topic. They should also be able to see how the highlighted words are particularly emotional and meant to attract the audience to the documentary. Adjectives demonstrate the use of literary choices that create an emotional connection, persuading people to watch and take seriously the issues discussed in the documentary. Before showing the film, make sure to emphasize that students should keep an open mind about the different ideas and information that they are about to see and hear. Again, it may also be helpful to warn students that there are scenes which show animals in distress.

While viewing the film, it may be helpful to have students work together to create reflective mind maps. Students should add words, ideas, pictures, feelings, and thoughts that they experience while watching the documentary and any other thoughts they have after.

This type of reflection is very important for students because it helps to consolidate ideas and also get students talking about the perspectives and feelings that they observe and experience during the documentary. Ask

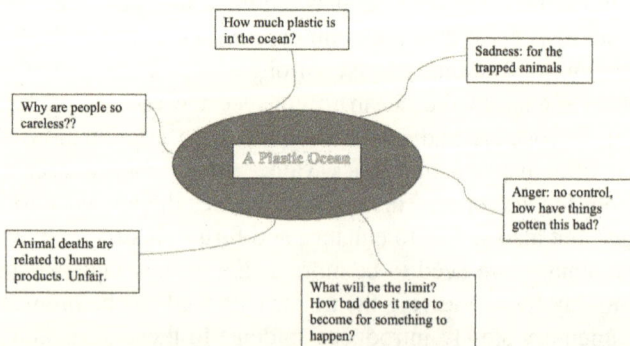

Figure 6.1. *A Plastic Ocean* Mind Map. Students should gather ideas about how the documentary has affected them and raised questions.
Source: Elke de Vries and Karen Ball

students to share their mind maps by posting them up around the room or in a hallway. Encourage students to consider what other groups have added.

Studying Persuasive Language

Ask students to describe what persuasive language is. From students' answers, write a definition together, asking students when they have seen or heard persuasive language being used. Make a list of examples and contexts from their ideas. Once students are engaged about what persuasive language is and when it is used, ask them to brainstorm how they could use it to encourage environmental awareness. These probing questions should direct students toward a better understanding about how language can be utilized to achieve a goal and what stories could be told in this effort.

A strategy we have used to check for understanding is to give examples of persuasive language and then ask students to match them to the technique that is used. Table 6.3. provides an example for the documentary that could be adapted for other contexts.

In demonstrating the persuasive techniques in the documentary, one can look firstly at the trailer as an introduction to not only topic but also the content. The trailer is 2:08 minutes long and highlights the main themes such as environmentalism, responsibility, and awareness, and it introducers viewers to the main producers of the movie.[34] As a short clip it is highly emotional and is meant to trigger the conscious of the viewers. Specifically, in the opening thirty seconds we see the emotional connection of the main director and producer, Craig Leeson, to the topic. This connects the viewer to Leeson's point

Table 6.1. Vocabulary Matching Game. Students should practice their vocabulary and match the word with the example.

Evidence	We are in trouble. Pollution means trouble. Plastic is causing us trouble.
Rhetorical Questions	If we continue to allow more plastic to be put into the ocean, then soon it won't be habitable.
Connotations	We need to make the first step; it is a problem we can solve together.
Repetition	We can see that over 60 percent of all plastics are not recycled, and from that, 30 percent ends up in the ocean.
Analogy	Where will the fish go when their homes are clogged with rubbish?
Formal Language	Nonbiodegradable plastics are a huge problem in ecosystems.
Cause-and-Effect Reasoning	The ocean looks like a floating plastic dump.
Inclusive Language	*The word* pollution *scares me as I know how wide-ranging that problem can be.*

of view and his passionate link to the topic, especially his early encounters with plastic pollution. The trailer can be a great example of persuasive techniques as it not only looks at the key points of the documentary itself, but it also uses persuasive techniques to ensure that the viewer watches/purchases the documentary itself.

The trailer utilizes the use of provocative scenes of plastic garbage to shock the viewer into being interested in the topic and the documentary itself. Specifically, the viewer sees scenes of children playing in garbage dumps, animals caught in plastic garbage, dead animals, and so on. These images are supposed to trigger an emotional response in the viewer and encourage them to learn more about plastic pollution. Not only these images, but Leeson introduces the idea that people have been eating plastic. This sensational idea hooks the viewer and entices them to watch further as the audience want to find out what potentially is happening to "our food chain" (1:13 minutes). This is then segmented with facts about plastic waste such as "every year 8 million tons of plastic waste are dumped into our oceans" (0:19 minutes). The use of the cause-and-effect technique is to develop the ideas that this is a much more important issue than most people consider. It also serves to highlight the legitimacy of the work of the scientists and reporters in the documentary. It is a logical approach to convince the audience of the reality of the problem and make it appear more believable.

The trailer also uses a wide range of evidence (something that is quite common in documentaries) which is used again to highlight the importance of the topic in that the consequences are catastrophic. The three main types of evidence that can be given are anecdotal, expert, and statistical. Three examples of this include:

- 0:55—Expert: examples of plastic that will never be able to be recycled, and to highlight the small proportion of plastic that actual is recycled.
- 1:04—Statistical: how many animals are affected by plastic and that pieces of plastic are turning up inside animals.
- 1:30—Anecdotal: that many businesses are only using plastic for food containers/wrappers. Few alternatives are being used.

Expert evidence makes the creators' position appear more creditable and reliable; in this example, we have many different scientists and environmental activists talking about the types of plastic found in the ocean and the realistic interpretation of how it will be recycled, if at all. Statistical evidence helps to make arguments seem more conclusive and more creditable. In this case, researchers talk about the amount of plastic that has been found in animals all around the world. Although some of the information can be taken out of context, the arguments presented in the trailer do highlight the seriousness of the topic. The anecdotal evidence is also used to establish a connection with the audience and for Leeson to appear more creditable to the audience. It uses his experience as an environmental activist and film maker to include real-life examples. Together, these different forms of evidence ensure that the viewer has a good overview of the different topics, but also entices them to watch the documentary to receive more information and get a better understanding of the topic as a whole.

Overall, this trailer was well made and demonstrates a well-thought-out attempt to not only entice viewers to go out and watch the documentary, but also to raise awareness on the topic of plastic pollution in oceans. We provide additional resources and strategies for studying persuasive language and techniques in the accompanying lesson plans.

CONCLUSION

Although plastic pollution is not specific to climate change, it is used as a tangible lens through which students might comprehend their carbon footprint and its impact on climate change. We feel this chapter will show students what they can do right now to confront global issues related to plastic pollution as well as to reduce their carbon footprint with a few small changes.

By using specific literary lenses such as poetry, short stories, and documentaries to learn about persuasive language teachers support and encourage students to be purposeful about their concerns and ideas that often fall outside of the classroom. Creating lessons that are related to real-life dire issues and that also use creative tools is a wonderful way to inspire the activism that we desperately need from our brilliant students, who will be the future leaders. With a variety of exciting sources and creative methods of teaching designed to reach all types of learners, including differentiated activities, we hope this chapter will help educators convey the significant impact of plastic pollution and empower students to take action against it. We wish you success in teaching our future leaders about the importance of language and literature in making change in their communities and world.

NOTES

1. "Youth for Climate Action," UNICEF, last modified November 16, 2021, https://www.unicef.org/environment-and-climate-change/youth-action.

2. George Michelsen Foy, "Humans Can't Plan Long-Term, and Here's Why," *Psychology Today*, last modified June 25, 2018, https://www.psychologytoday.com/us/blog/shut-and-listen/201806/humans-cant-plan-long-term-and-heres-why.

3. "Second Ocean Heroes Bootcamp Empowers Youth to—ProQuest," ProQuest, last modified 2019, www.proquest.com/docview/2247755857?parentSessionId =nHYJlc68znvLiO%2BnUAiyuH1ewCtgWl3hUP1sBKCH1rQ%3D&pq-origsite =primo&accountid=14682 (Accessed 18 June 2022).

4. Pro Quest, "Second Ocean Heroes Bootcamp Empowers Youth to—ProQuest."

5. Pro Quest, "Second Ocean Heroes Bootcamp Empowers Youth to—ProQuest."

6. Pro Quest, "Second Ocean Heroes Bootcamp Empowers Youth to—ProQuest."

7. Maggie Astor, "7 Times in History When Students Turned to Activism," *The New York Times*, last modified March 5, 2018, https://www.nytimes.com/2018/03/05 /us/student-protest-movements.html

8. Charles Moore, "Plastic Pollution," *Encyclopedia Britannica*. https://www .britannica.com/science/plastic-pollution, (accessed February 22, 2021)

9. "The Lifecycle of Plastics," WWF, last modified June 19, 2018, https://www .wwf.org.au/news/blogs/the-lifecycle-of-plastics#gs.thymxr.

10. Perry Wheeler and Ivy Schlegel, "Preventing Plastic Pollution," Greenpeace USA, Last modified May 1, 2020, https://www.greenpeace.org/usa/oceans/preventing -plastic-pollution/.

11. Laura Parker, "Plastic Pollution Facts and Information," *National Geographic*. last modified June 7, 2019, https://www.nationalgeographic.com/environment/article /plastic-pollution.

12. Brian Clark Howard, Sarah Gibbens, Elaina Zachos, and Laura Parker, "A Running List of Action on Plastic Pollution," *National Geographic*, last modified June 10, 2019, https://www.nationalgeographic.com/environment/article/ocean-plastic -pollution-solutions.

13. "Japanese Schools Use Cleanups to Teach Plastic Pollution," *Earth Day*, last modified September 18, 2019, www.earthday.org/students-in-japan-use-cleanups-to -educate-about-plastic-pollution/,(Accessed 18 June 2022).

14. "How Plastic Pollution Is Affecting Seals and Other Marine Life," World Animal Protection, last modified September 9, 2020, https://www.worldanimalprotection .org/news/how-plastic-pollution-affecting-seals-and-other-marine-life

15. World Animal Protection, "How Plastic Pollution Is Affecting Seals and Other Marine Life."

16. World Animal Protection, "How Plastic Pollution Is Affecting Seals and Other Marine Life."

17. World Animal Protection, "How Plastic Pollution Is Affecting Seals and Other Marine Life."

18. "100 Ocean Plastic Pollution Statistics & Facts," *It's a Fish Thing*, last modified May 11, 2022, https://www.itsafishthing.com/marine-ocean-plastic-pollution-

19. "You Can Help Prevent Plastic from Harming Marine Wildlife," Seattle Aquarium, https://www.seattleaquarium.org/blog/you-can-help-prevent-plastic-harming -marine-wildlife, (Accessed June 18, 2022).

20. *It's a Fish Thing*, "100 Ocean Plastic Pollution Statistics & Facts."

21. Derek Gladwin, "Ecocriticism," Oxford Bibliographies Online Datasets, last modified 2017, p. 1, https://doi.org/10.1093/obo/9780190221911-0014.

22. Emily Drabble, "What Are the Best Eco Books for Children and Teens?" *The Guardian*, Guardian News and Media, last modified April 20, 2015. https://www .theguardian.com/childrens-books-site/2015/apr/20/climate-change-best-eco-books -for-children-and-teens.

23. Naomi Shihab Nye, "'World of the Future, We Thirsted,'" *The New Yorker*, last modified August 5, 2019, https://www.newyorker.com/magazine/2019/07/29/world -of-the-future-we-thirsted.

24. Kelly Roper, "Poems about Pollution," LoveToKnow Corp., https://greenliving .lovetoknow.com/Poems_About_Pollution, (Accessed January 12, 2021).

25. "Fridays for Future," Fridays for Future, https://fridaysforfuture.org/ (accessed January 12 2020).

26. Fridays for Future, "Fridays for Future."

27. "About a Plastic Ocean," Plastic Oceans International, last modified October 8, 2021, https://plasticoceans.org/about-a-plastic-ocean/.

28. "Rethink Plastic. Save Our Seas • Plastic Oceans International," Plastic Oceans International, last modified January 5, 2021, https://plasticoceans.org/.

29. "Resources—Info," Plastic Oceans International, last modified October 22, 2020, https://plasticoceans.org/plastic-pollution-info-resources/.

30. "Poetry and the Environment," *Poetry Foundation*, https://www.poetryfoundation .org/collections/146462/poetry-and-the-environment, (accessed June 10, 2022).

31. Kelly Roper, "Poems about Pollution," LoveToKnow Corp., https://greenliving .lovetoknow.com/Poems_About_Pollution, (accessed January 12, 2021).

32. Laura Parker, "Plastic Pollution Facts and Information," *National Geographic*, Last modified June 7, 2019, https://www.nationalgeographic.com/environment/article /plastic-pollution.

33. "About a Plastic Ocean," Plastic Oceans International, last modified October 8, 2021, https://plasticoceans.org/about-a-plastic-ocean/.

34. Plastic Oceans International, "About a Plastic Ocean."

BIBLIOGRAPHY

"100 Ocean Plastic Pollution Statistics & Facts." *It's a Fish Thing*. Last modified May 11, 2022. https://www.itsafishthing.com/marine-ocean-plastic-pollution-

"About A Plastic Ocean." Plastic Oceans International. Last modified October 8, 2021. https://plasticoceans.org/about-a-plastic-ocean/.

Astor, Maggie. "7 Times in History When Students Turned to Activism." *The New York Times*. Last modified March 5, 2018. https://www.nytimes.com/2018/03/05/ us/student-protest-movements.html.

Drabble, Emily. "What Are the Best Eco Books for Children and Teens?" *The Guardian*. Guardian News and Media. Last modified April 20, 2015. https:// www.theguardian.com/childrens-books-site/2015/apr/20/climate-change-best-eco -books-for-children-and-teens.

Foy, George Michelsen. "Humans Can't Plan Long-Term, and Here's Why." *Psychology Today*. Last modified June 25, 2018. https://www.psychologytoday .com/us/blog/shut-and-listen/201806/humans-cant-plan-long-term-and-heres-why.

"Fridays for Future." Fridays for Future. https://fridaysforfuture.org/ (Accessed January 12, 2020).

Gladwin, Derek. "Ecocriticism." Oxford Bibliographies Online Datasets. Last modified 2017, p. 1. https://doi.org/10.1093/obo/9780190221911-0014.

Howard, Brian Clark, Sarah Gibbens, Elaina Zachos, and Laura Parker. "A Running List of Action on Plastic Pollution." *National Geographic*. Last modified June 10, 2019. https://www.nationalgeographic.com/environment/article/ocean-plastic -pollution-solutions.

"Japanese Schools Use Cleanups to Teach Plastic Pollution." *Earth Day*. Last modified September 18, 2019. www.earthday.org/students-in-japan-use-cleanups-to -educate-about-plastic-pollution/. (Accessed 18 June 2022).

Moore, Charles. "Plastic Pollution." *Encyclopedia Britannica*. https://www.britannica .com/science/plastic-pollution. (Accessed February 22, 2021).

Nye, Naomi Shihab. "'World of the Future, We Thirsted.'" *The New Yorker*. Last modified August 5, 2019. https://www.newyorker.com/magazine/2019/07/29/ world-of-the-future-we-thirsted.

Parker, Laura. "Plastic Pollution Facts and Information." *National Geographic*. Last modified June 7, 2019. https://www.nationalgeographic.com/environment/article/ plastic-pollution.

"Poetry and the Environment." *Poetry Foundation*. https://www.poetryfoundation.org /collections/146462/poetry-and-the-environment. (Accessed June 10, 2022)

"Resources—Info." Plastic Oceans International. Last modified October 22, 2020. https://plasticoceans.org/plastic-pollution-info-resources/

"Rethink Plastic. Save Our Seas • Plastic Oceans International." Plastic Oceans International. Last modified January 5, 2021. https://plasticoceans.org/.

Roper, Kelly. "Poems about Pollution." LoveToKnow Corp. https://greenliving .lovetoknow.com/Poems_About_Pollution. (Accessed January 12, 2021).

"Second Ocean Heroes Bootcamp Empowers Youth to—ProQuest," ProQuest. Last modified 2019. www.proquest.com/docview/2247755857?parentSessionId =nHYJlc68znvLiO%2BnUAiyuH1ewCtgWl3hUP1sBKCH1rQ%3D&pq-origsite =primo&accountid=14682 (Accessed 18 June 2022).

"The Lifecycle of Plastics." WWF. Last modified June 19, 2018. https://www.wwf .org.au/news/blogs/the-lifecycle-of-plastics#gs.thymxr.

Wheeler, Perry, and Ivy Schlegel. "Preventing Plastic Pollution." Greenpeace USA. Last modified May 1, 2020. https://www.greenpeace.org/usa/oceans/preventing -plastic-pollution/.

"You Can Help Prevent Plastic from Harming Marine Wildlife." Seattle Aquarium. https://www.seattleaquarium.org/blog/you-can-help-prevent-plastic-harming -marine-wildlife. (Accessed June 18, 2022).

"Youth for Climate Action." UNICEF. Last modified November 16, 2021. https:// www.unicef.org/environment-and-climate-change/youth-action.

Chapter Seven

Ecohorror, Terrorism, and Inadequate Representation of Global Warming in M. Night Shyamalan's *The Happening*

Tatiana Konrad

THE NEW HORROR

The genre of horror is in many ways interesting for filmmakers. Due to its peculiarities, horror has always been an outstanding medium to explore multiple political and sociocultural issues. On-screen violence and psychological tension, two of the genre's key characteristics, have been used as tools to mirror problems specific to these films' respective eras. From the 1974 *Texas Chainsaw Massacre*'s commentary on the US involvement in the Vietnam War to the many torture-porn horrors that criticized the inhumane violence of some American soldiers at Abu Ghraib—of which *Saw* (2003) and its sequels are perhaps the best known—the horror film has proved to be a powerful medium to discuss certain events, to spread awareness about pressing domestic and international problems that emerge, and to display social and political anxieties.

Horror films have been undergoing numerous transformations. Peaking in quality and popularity during the 1970s and 1980s, these films experienced a notable downturn in the 1990s, a slide that continued until recently, with horror films being, according to Steffen Hantke, now "just aren't any good."[1] Examples like Jordan Peele's *Get Out* (2017), however, clearly indicate that horror experiences a revival today. Whatever diminishment there may have been in plot, acting, the authenticity of what is shown, and the story's influence on the viewer, the ability of horror films to speak about serious issues doubtless remains as powerful as it has always been. It certainly possessed such power in the aftermath of the terrorist attacks of September 11, 2001. Just like other literary and cinematic genres, horror was informed by the attacks and was itself transformed in their wake. Anna Froula contends: "The aftermaths of 9/11 have compounded chaos in warzones and increased national and international surveillance. They intersect with crises in climate change and infrastructure readiness (or lack of readiness), global hunger, increasingly hypertoxic environments, and 'superbugs,' such as the Zika virus. Such intersections suggest a wider scope of 'post-9/11 texts.'"[2] Indeed, war as a political event and climate change as a global issue have been some of the most striking problems that horror films have been dealing with since 9/11. This chapter will demonstrate what happens when these two concerns are merged in one film, and climate change becomes a metaphoric projection of the most famous war of the twenty-first century—the War on Terror. Dealing with a unique type of ecohorror, this chapter examines M. Night Shyamalan's film *The Happening* (2008). Zeroing in on cases of mass suicide taking place throughout the United States that were provoked by plant-secreting toxins, *The Happening* compares global warming to a terrorist attack. Through its focus on New York, the citizens who commit suicide by jumping off roofs, the media that spread the news about the incidents, as well as the armed survivors who protect their property, the film reimagines the tragic events of 9/11 and the War on Terror that followed, so that that particular tragedy becomes a way to understand the problem of climate change. Here in this chapter, I demonstrate the limits and the danger of portraying climate change as a problem viewed through the filter of 9/11 and its ramifications; yet I also argue that it is through narratives like *The Happening* that audiences worldwide become informed about the tragedy and the scale of the largest environmental issue that humanity faces today, whose existential threat we cannot ignore.

Through its multiple ecological monsters, horror films can misrepresent and, in a way, even mock environmental transformation that is taking place. Consider the portrayals of monstrous nature in, for example, Gordon Douglas' *Them!* (1954), Alfred Hitchcock's *The Birds* (1963), Mark Atkins' *6-Headed Shark Attack* (2018), and the series *The Strain* (2014–2017).[3] The Nature that the viewer sees in these films and series differs considerably from Nature as we observe it in reality. Yet such portrayals do not claim to be reality. They force the viewer to imagine Nature anew, not as peaceful, nurturing, and giving, but rather as dangerous, hostile, and unrecognizable. What if one day such harmless (as humanity tends to think) organisms as plants will be able to attack us? Imagining various horrifying scenarios in which Nature and the environment are no longer welcoming, filmmakers and writers of horror witness the environmental crisis that is taking place and, through their texts, turn their viewers and readers into witnesses, too. They "question the story of our time" and retell it in a new way to fully unveil "its meaning, submerged beneath public ideology, discourse and action."[4] Envisioning climate change through fiction is a complex process. Many people do not even notice climate change and cannot estimate its danger in real life. As a result, fictional representations of climate change fail to communicate the full and accurate scale of the problem. Amitav Ghosh explores the problem of *"[i]mprobability"* of fictional portrayals of climate change and argues that artists have not done enough to present the challenges of climate change in their art, specifically fiction.[5] While Ghosh's observation is correct, I claim that there is an educational value in the fictional narratives that from a scientific viewpoint distort the meaning of climate change. Horror films and literature, for example, emphasize the dreadful ramifications of climate change by imagining the planet and its inhabitants being largely transformed due to environmental degradation. These narratives, thus, directly address the problem that Naomi Klein formulates as follows: "Climate change is slow, and we are fast. When you are racing through a rural landscape on a bullet train, it looks as if everything you are passing is standing still: people, tractors, cars on country roads. They aren't, of course. They are moving, but at a speed so slow compared with the train that they appear static."[6] Horror can hence be an effective genre to examine environmental degradation, for it fast-forwards our reality to the moment when no changes can be made, thus reinforcing the horror of both our procrastination and inability to sufficiently address climate crisis and the result of that procrastination. While it might be problematic to deal with horror films in the classroom due to their graphic content, educators can consider including clips or stills from selected films to illustrate certain issues.

THE WAR ON TERROR IN A POST-ECOLOGICAL REALM

The depictions of the War on Terror in film are numerous. The infamous US military interventions in Afghanistan and later in Iraq that were direct responses to the violence committed on American soil have drawn the attention of many filmmakers. From Robert Redford's *Lions for Lambs* (2007) to Peter Berg's *Lone Survivor* (2013) to Kathryn Bigelow's *The Hurt Locker* (2008) and *Zero Dark Thirty* (2012) to Clint Eastwood's *American Sniper* (2014) and beyond, there are manifold cinematic examples that discuss the role of the United States in the two wars. These films have also influenced, or rather formed, a cultural perception of the victim and the enemy, revealing the imperfections within US policies, or, on the contrary, stigmatizing certain religious and national groups in the eyes of some. In this regard, Terence McSweeney laments: "Throughout its history American cinema has rarely offered sympathetic images of the Other, those figures who do not correspond to what a society defines as its 'norm' whether in terms of race, nationality, gender, or sexuality. A study of post-9/11 American film sees a continuation of this practice of failing to recognise or portray the essential humanity of alternate lives."[7] This "cinema simplicity"[8] that so blatantly facilitates a division into good and bad characters, heroes and villains, has proved to be effective in communicating certain messages to the viewer, yet in a more global construction of the meaning of the War on Terror it has been wrong. Nonetheless, these films have created such a solid image of terrorism as the predominant evil for everyone in the twenty-first century that no other portrayal will conceivably be able to overthrow it and create a new cultural perception of evil, at least in the near future. Terrorism is such an obvious lens for evil because it is incredibly tangible and immediate. Climate change, on the other hand, is a far more abstract threat even though its consequences are so severe. Filmmakers are appealing to neuroscience, knowing that viewers will react more affectedly to the very real threat of terror than to the more undefinable that is climate change. It is thus unsurprising that terrorism has been frequently used and in some cases—paradoxically—even abused by filmmakers to portray evil at its utmost.

Outlining some of the key events depicted in horror films after 9/11, Abiva Briefel and Sam J. Miller mention "the destruction of the World Trade Center, the Iraq War, and the tortures perpetuated at Abu Ghraib and other detention centers."[9] 9/11 and the War on Terror became the key occasions on which numerous directors centered the plots of their horror films. Yet it should be noted that "the memory of 9/11 can never be obliterated from the American national consciousness, no matter how crassly it is exploited for personal, political, or corporate gain."[10] But exploited in film 9/11 is. One such example that does it in a thoughtful and complex way is *The Happening*—a film that, combining the War on Terror and climate change, explores the more profound political debates behind the two issues, providing a cultural and environmental critique of the actions of humanity and thus bringing the two scenarios of violence into collision with each other. Through its portrayal of plants as terrorists and climate change as the War on Terror, the film closely engages with the problem of exploitation and Western supremacy that has become particularly visible in the relationship between the United States and the Middle East since the First Gulf War and only intensified after 9/11. Yet the plot of *The Happening* becomes a tool of environmental criticism when, in a manner akin to the portrayal of the complex relationship between the United States and the Middle East, it depicts the role of the West as exploitative toward nature. What does the audience make of the choice to display plants as terrorists?

While in the representations of terrorists the enemy is usually dehumanized, here monstrous plants become devitalized—a transformation that is enabled because of the destructive and rapacious actions of humans toward nature.

Briefel and Miller make a crucial observation: "We have come to expect that a monster is never just a monster, but rather a metaphor that translates real anxieties into more or less palatable forms. According to a 2005 article from the *Nation*, 'Every generation gets the movie monster it deserves.'"[11] Have we finally come to the time when we must face the real monster of today's world—climate change? But what do we do with the terrorist/plant-monster? My contention is that precisely because of the multilayered complexity of twenty-first-century political and environmental issues, we now experience a cultural confusion skillfully transferred onto the screen. It is because of our social bewilderment and an inability to solve the problems that have been neglected and thus have intensified for decades that humanity finds itself in a state of desperation and psychological collapse. The chaos within the current social, political, and ecological environments creates images of heroes, villains, and victims that are more ambiguous than ever before and can no longer be easily and intelligibly interpreted. That is why we have examples like *The Happening* that portray climate change as the War on Terror and plants as terrorists. These tropes are an expression of the despair, lack of knowledge, and uncertainty about what would be a clear plan to deal with climate change. Our overt inadequacy in understanding/portraying climate change scares global audiences and thus constructs today's true (eco)horror.

THE HAPPENING AND THE NEW DIRECTIONS IN PLANT HORROR

McSweeney detects an important tendency in the films about 9/11 and/or its aftermath, arguing that "American cinema has been reluctant to represent the terrorist attacks on September 11th 2001 directly on the screen."[12] The scholar makes a firm contention that numerous blockbusters focusing their plots on aliens invading the world "fill their screens with barely coded images and situations so self-consciously designed to evoke 9/11 and the war on terror."[13] It is through "their potent realisations of an alien Other" that these "remakes" of 9/11 communicate "the particular fears and fantasies that characterised the decade."[14]

The Other as a cultural concept has been transforming, meeting the needs of the genre and the filmmakers' primary purposes: Muslim terrorists have morphed into aliens and now, as this chapter argues, plants. Relying on Dawn Keetley's classification of plant horror, which outlines why plants can easily be turned into objects of horror, I draw parallels between cultural imaginings

of 9/11 and climate change. Keetley argues that in horror films plants represent *"absolute alterity"*[15]; they have "a unique ability to strike us blind"[16] and
"despite their rootedness in place, plants *do* move."[17] But, she adds, "maybe
we are also *like plants*,"[18] and "plants doubly threaten to stage a vengeful
return, forming a potent force of both the repressed *and* the oppressed."[19]
Finally, the rise of plants represents "an absolute rupture not only of normality (as always happens in horror), but of the entirety of the known world and
its fundamental structuring rules."[20] Through the themes of hostility, mutation, and vengeance, films like *The Happening* explore the new directions in
plant horror that are deeply informed by, first, the existing portrayals of trees
rising to a similar level of representation, of both knowledge and revenge,
as in J. R. R. Tolkien's *The Lord of the Rings* (1954) and its sequels, and,
second, the political and sociocultural concerns of the United States today, so
that the genre of ecohorror is transformed into a profoundly disturbing, political, and polemical story of global warming. In reaction to anthropocentrism,
the plants have agency in these depictions, which is an important reminder
that humans coexist with the nonhuman world. This, in turn, emphasizes the
complexity of environmental crisis, for it impacts not only humanity but also
more-than-humans. Giving agency to nonhumans such as plants, cultural
texts recognize the problematic nature of anthropocentrism and foreground
the importance of moving beyond human in order to understand and produc-
tively address environmental crisis, including climate change.

Adam Trexler's essay "Mediating Climate Change: Ecocriticism, Science
Studies, and *The Hungry Tide*" opens with the sentence: "Imagining climate change is an enormously difficult task."[21] This observation is arguably
crucial in the cultural production and interpretation of this environmental
issue—because, of course, neither the "cause" nor the "effects" can be clearly
detected, or, to borrow from Trexler, "[t]here are no clear villains."[22] The
situation even worsens when we start thinking of climate change as a "pre-
dominantly scientific problem," which results, according to Ursula Kluwick,
in two major problems: first, one tends to misinterpret that climate change's
"intricacies, its complexities, and its logic are [. . .] only accessible to the
select few," and, second, "in engaging with climate change, we have tended
to concentrate on the search for scientific, technological, and, increasingly,
economic solutions."[23] The way we have globally been interpreting climate
crisis is thus largely insufficient and frequently misleading.

Scholars have already mentioned the probable connections of what is
occurring in *The Happening* to 9/11. Jericho Williams, for example, claims
that even the film's trailer makes it clear that the city's misfortunes are due
to either "terrorists" or "aliens."[24] Williams elucidates this simplification of
the enemy, the inability to develop a properly formulated pattern about a new
type of enemy, and an atmosphere of true ecological calamity as follows: "By

placing an emphasis on fears of terrorist attacks, chemical contamination, or extraterrestrial invasions, the trailer appeals to reviewers' awareness of catastrophes in current events and of the familiar tropes of alien invasions, rather than revealing the monstrous force as potentially plant-related."[25] This might be the filmmaker's calculated decision given his appreciation of viewers' sensibilities. But this is also a failure to directly speak about environmental degradation as a serious threat to the life on Earth. Familiarity with the enemy and a sense of expectation of the horror—despite how paradoxically this sounds—appears to be key in the creation of a disastrous environment and the presumption of a nation under threat. It is through the introduction of already known patterns that the film "invite(s) us to reconsider the symbiosis of nature and humans in a future where the natural world is . . . undergoing disturbing dramatic changes."[26] Froula openly characterizes the plot in *The Happening* as being similar to 9/11 cinematic discourses in the way that it "imagine(s) human-created natural and social collapse."[27] The explanation that *The Happening* provides its viewers with regarding the emerging chaos and a possible threat to the existence of New Yorkers, and later to global humanity, is that it is the danger that comes from the outside—from the evil Other.

While the inclusion of the discourse of Otherness seems a rather predictable choice, it is, as this chapter contends, inadequate to the representation of climate change, primarily because it creates a distorted cultural perception of the causes and effects of the environmental crisis. The references to 9/11 help outline not only the event's "pretrauma"[28]—a term I borrow from E. Ann Kaplan—but also, crucially, its *post*trauma, for by the time the film had been released, the American and the global viewers had become aware of the after-effects of 9/11: the American nation continues to remember the terrorist attacks' innocent victims but the world in general still feels its insecurity and vulnerability with regard to the faceless enemy—terrorists. The overt comparison of climate change to 9/11 is neither correct nor wrong; it is simply inadequate. But let us first examine the references to 9/11 and the War on Terror in *The Happening* in greater detail, making use of Keetley's classification.

New York City as the locale of the first "attacks'"; the multiple suicides, particularly committed by people jumping from rooftops; the unawareness of citizens caught by surprise by the attacks, and the quickly ensuing panic—all these details serve to recreate the tragedy of September 11, 2001, in its most enduring memories. The film then focuses on a specific family (although never allowing the viewer to doubt that all other families have found themselves in similar situations), that of the high school science teacher Elliot Moore (Mark Wahlberg) and his wife, Alma Moore (Zooey Deschanel), who, together with their friend Julian (John Leguizamo) and his young daughter

Jess (Ashlyn Sanchez), hurriedly leave New York City hoping to escape the calamity. When, later, Elliot, Alma, and Jess are alone, Julian having left them in an attempt to find his wife, the film only intensifies the idea of an American family being targeted by an evil entity and appearing innocent and hopeless before the threat. Over the course of the film, the family discovers that plants are responsible for the unfolding events due to their production of toxins that, as the characters eventually find out, can influence only larger groups of people. Later on, the toxins also become dangerous for individuals and the family decides to part ways. Alma remains with Jess (thus embodying a cultural image of mother and child—the most vulnerable group in the film) in one part of a house that they have found on the way, whereas Elliot stays in the basement. The separation of the man from his family can be metaphorically correlated with the separation that thousands of American families experienced when their family members joined the military and were deployed to Afghanistan and Iraq. The only way that Elliot can communicate with Alma and Jess is through a talking tube. Elliot's ultimate expression of love for his wife and his desire to spend time with her rather than in the basement, even if it kills him, serves to intensify the longing for seeing family and the inability to live through the separation that many military families and their relatives and friends have experienced during the War on Terror, and specifically during the two wars in the Middle East. Finally, there is another scene that serves to support the idea that the War on Terror is here a symbol of climate change. When Elliot, together with the two teenage boys that he has met during the journey, finds a house and asks their inhabitants to let them in and thus safeguard them from the dread influence of toxins, the people in the house respond aggressively to this request, saying that they will protect their property from encroachment by any strangers. This desire to protect the home (whether understood as an individual home in the United States or as the homeland of the United States) from any invasion overtly hints at what the War on Terror is about: the aims of never allowing anyone to attack the country again, of protecting its citizens, and of fighting against any intruders. And the residents do fight: the two boys are shot from inside the house.

What role do plants play in this story of 9/11, the War on Terror, and climate change? They are the objects of horror and thus are viewed as the enemy. Keetley singles out six main criteria that explain the horrifying nature of plants. These criteria, I argue, help one juxtapose plants and terrorists and compare them to each other, illuminating the metaphor used to characterize the enemy in post-9/11 horror films. First, the "absolute alterity" of the plants consists in their "[n]onhuman, nonanimal"[29] nature. A similar interpretation of terrorists has been formulated within cultural discourses that stress the Otherness—cultural, linguistic, and so on—of terrorists, their truly nonhuman nature. Second, the "ability to strike us blind" is another characteristic that is easily associated with terrorists: indeed, none of the terrorist attacks that have happened were known about (or such knowledge always came too late to prevent an attack). Third, the fact that plants have roots does not stop them from moving to other places. Having training bases in the Middle East and Central Asia, terrorists have, indeed, proved to be very mobile, which, in some way, explains the temporal and geographical unexpectedness of all attacks. Fourth, a certain similarity between humans and plants on the basis of chemistry and biology can be compared to a kind of similarity between terrorists and nonterrorists/(potential) victims in the way that, despite the cultural and ideological otherness, there is biological sameness among us all. Natania Meeker and Antónia Szabari's contention is striking here: "while plants have nothing of the human within them, humans all contain something of the vegetal order."[30] To rephrase this: terrorists have no humanness or goodness within them, yet nonterrorists can generate violence and evil (after all, nobody is born a terrorist). Fifth, the avenging nature of plants can be similar to that of certain terrorists: going back to 9/11 again, the attacks have been frequently interpreted as a response to the presence of US military troops in Saudi Arabia and the too active role that the "infidels" played in certain Middle Eastern affairs. Finally, the invasion of plants is in itself an "absolute rupture not only of normality . . . , but of the entirety of the known world and its fundamental structuring rules": this can be equated to the way the 9/11 terrorist attacks shattered George H. W. Bush's idea of a post–Cold War "new world order."

This imagining of climate change certainly reveals new ways to interpret nature and the environmental crisis in plant horror in general and in post-9/11 ecohorror in particular. According to Kaplan, in *The Happening* "climate change is envisaged as uncanny—in some way beyond rationality or understanding."[31] In turn, Williams comments on the choice to portray the enemy as a plant: on the one hand, the film reminds us that "the entire ecosystem is continually present within us in the form of the air we breathe," and thus "[n]o matter who we are or where we go, we all bear responsibility for the degradation of the Earth because our bodies react to and depend on the conditions that we create,"[32] but on the other hand, it "defies the conventional post-Enlightenment notion that Western societies have mastered the natural world."[33] Human responsibility is, hence, in a sense put aside, although the film insistently reminds us that neither geographically nor temporally can humans escape the dangers that the modified-by-anthropogenic-factors environment poses to humanity. To borrow from Williams: "The film's horror stems from the idea that human behavior may provoke deadly environmental reactions that are neither foreseeable nor easily comprehensible; in the direst of these situations, there may not be enough time for an adequate response to prevent a massive loss of human lives, a frightening proposition in light of real-world issues such as global warming and human-generated pollution and toxicity."[34] The film thus suggests that the transformations that are happening and will continue to happen to our planet are so dramatic that we neither can fully realize the scale of the problem nor will we be able to deal with the dreadful ramifications if we continue to think that climate crisis is not an urgent issue.

The post-9/11 (eco)horror is thus modified according to the issues that are urgent today. Whether such modifications are justifiable is another question, but they undoubtedly reveal the chaos within contemporary political/military and environmental thinking. Numerous films and TV series attempt to portray negligent human attitudes toward the environment and the results of this carelessness, introducing parasites and monsters that live in polluted lands and waters. From the series *The Strain* (2014–2017) to Guillermo del Toro's recent Academy Award–winning *The Shape of Water* (2017)—all these cultural products make us think of the mutations and metamorphoses facilitated by the new environmental reality, introducing them almost as a norm. This is what the new ecohorror is about—the inability to control or predict transformations, pacify nature, and suppress environmental collapse.

The post-9/11 ecohorror also reveals an important characteristic of today's attitude toward the largest environmental problem—climate change. To speak more specifically about this attitude, human powerlessness is at the center of many recent films, including *The Happening*. The combination of environmentalism with a representation of militarism opens another significant perspective on our understanding of climate change and its cultural perception. The metaphorical reference to 9/11 in *The Happening*, indeed, by no means undermines or belittles the trauma caused by the terrorist attacks; on the contrary, it reveals the enduring psychological effects of that tragic day on the United States as a nation and the deep cultural disturbance that it has generated worldwide. Yet the choice to talk about climate change through the prism of 9/11 is, to put the matter as simply as possible, dangerous. On the one hand, familiar plots and cinematic/narrative patterns help the audience better understand what is being shown, encoding the story about climate change as one that is traumatic, dangerous, and life-changing. On the other hand, the ideas of militarism, revenge, unjustness, and aggressiveness that they bring along become, in many ways, perverse in a representation of anthropogenic climate change. Portraying plants as terrorists inevitably foregrounds the issue of nature's aggressiveness toward humanity, which *a priori* is wrong, for humans have caused the contemporary predicament of climate change. Presenting the fight against climate change as akin to the War on Terror enables a distorted understanding of the role humans play in the current environmental crisis: Are we fighting against nature? *The Happening* hardly presents this view; rather, it claims that we are fighting against the transformations that are happening in the natural world (evil plants here figuratively stand for pollution and mutated natural beings). Yet this ambiguity between "nature as the victim" and "nature as the enemy" that *The Happening* foregrounds does not help clearly position the human within these new discourses of ecohorror. During our time of climate change, nature seems to be both a victim and a villain, and the militarism that the theme of 9/11 brings into the narrative only underlines the hostility of humans—which existed when the natural world was systematically and steadily deteriorated and in part destroyed by humans, and which, it seems, will continue to persist as climate change threatens human existence and thus in a way can be interpreted as an enemy. There is one way, however, that the choice to speak about the potential battle against climate change as analogous to the War on Terror can be interpreted as beneficial. The aggressiveness of the United States toward the Middle East and its invasions of Afghanistan and Iraq proved to be ineffective in fighting against terrorism. One, therefore, can argue that starting another "War on Terror," that is, being hostile against nature and continuing to demonstrate superiority over it, will be just as ineffective and will provoke only more chaos. Through the scenarios of 9/11 and the War on Terror, *The Happening* reveals the (eco)horror of the current situation regarding climate change, the unwillingness of some to believe in its reality, and the inability of many to understand that climate change can eventually result in a fight for survival.

CONCLUSION

The calamity of September 11, 2001, is regarded as one of the biggest tragedies in US history. The two military interventions in Afghanistan and Iraq that followed the attacks became known as the most controversial wars in modern history. The War on Terror, while being justifiable and necessary, has proved to be endless and effective only in part. All these events are historically, politically, and culturally important, yet complex. Combining them with other serious issues like climate change—through either cultural portrayals or the drawing of theoretical interpretations—on the one hand, indeed, pushes cultural consciousness to the border where the human mind finally treats environmental problems with the seriousness and urgency they deserve. Yet, on the other hand, the choice to do so, like in *The Happening*, can hardly be called ideal. Such a portrayal doubtless demonstrates the cultural confusion of our post-9/11 world; yet it fails to clearly recognize that climate change is a real problem that is only getting worse. The sudden cessation of the plant attack in the United States toward the end of the film not only undermines any sense of the danger of climate change, but it also seems to suggest that humans do not need to do anything for climate change to stop. The film rehabilitates itself only in the final scene, which depicts a sudden attack by toxic plants in France, thus accentuating the global scale of climate change and putting forward the idea that humanity will be able to solve the existing environmental issues only collectively, when countries stop talking primarily about national environmental policies and instead introduce global rules to save the planet. By using clips and stills from *The Happening*, educators can engage with their students about climate change, discussing the advantages and disadvantages of portraying climate crisis as a horror story, analyzing the images of monstrous nature (and monstrous plants, in particular) in the film, examining the filmmaker's choice to envision climate change as a national tragedy through 9/11, and outlining potential solutions to climate crisis as, indeed, a global issue.

NOTES

1. Steffen Hantke, "Academic Film Criticism, the Rhetoric of Crisis, and the Current State of American Horror Cinema: Thoughts on Canonicity and Academic Anxiety," *College Literature* 34, no. 4 (2007): 191, accessed February 19, 2018, http://www.jstor.org/stable/25115464.

2. Anna Froula, "What Keeps Me Up at Night: Media Studies Fifteen Years after 9/11," *Cinema Journal* 56, no. 1 (2016): 113, accessed February 19, 2018, https://doi.org/10.1353/cj.2016.0056.

3. For more on the portrayals of monstrous nature in *The Strain*, see Tatiana Prorokova-Konrad, "The Ecohorror of *The Strain*: Plant Vampires and Climate Change as a Holocaust," in *The Global Vampire: Essays on the Undead in Popular Culture Around the World*, ed. Cait Coker (Jefferson NC: McFarland, 2020), 189–99.

4. Nadine Gordimer, "Witness: The Inward Testimony," Government of India, Ministry of External Affairs, November 10, 2008, accessed January 20, 2021, https://www.mea.gov.in/Speeches-Statements.htm?dtl/1754/WITNESS++THE+INWARD+TESTI.

5. Amitav Ghosh, *The Great Derangement: Climate Change and the Unthinkable* (Chicago: University of Chicago Press, 2016), 16; italics in original.

6. Naomi Klein qtd. in Timothy Morton, *Dark Ecology: For a Logic of Future Coexistence* (New York: Columbia University Press, 2016), 25.

7. Terence McSweeney, *The "War on Terror" and American Film: 9/11 Frames per Second* (Edinburgh: Edinburgh University Press, 2014), 33.

8. John Markert, *Post-9/11 Cinema: Through a Lens Darkly* (Lanham: Scarecrow Press, 2011), 1.

9. Aviva Briefel and Sam J. Miller, introduction to *Horror after 9/11: World of Fear, Cinema of Terror*, ed. Aviva Briefel and Sam J. Miller (Austin: University of Texas Press, 2011), 1.

10. Wheeler Winston Dixon, "Introduction: Something Lost—Film after 9/11," in *Film and Television after 9/11*, ed. Wheeler Winston Dixon (Carbondale: Southern Illinois University Press, 2004), 3.

11. Briefel and Miller, introduction to *Horror after 9/11*, 4.

12. McSweeney, *The "War on Terror" and American Film*, 135–36.

13. McSweeney, *The "War on Terror" and American Film*, 136.

14. McSweeney, *The "War on Terror" and American Film*, 136–37.

15. Dawn Keetley, "Introduction: Six Theses on Plant Horror; or, Why Are Plants Horrifying?," in *Plant Horror: Approaches to the Monstrous Vegetal in Fiction and Film*, ed. Dawn Keetley and Angela Tenga (London: Palgrave Macmillan, 2016), 6; italics in original.

16. Keetley, "Introduction," 10.

17. Keetley, "Introduction," 13; italics in original.

18. Keetley, "Introduction," 16; italics in original.

19. Keetley, "Introduction," 19; italics in original.

20. Keetley, "Introduction," 22.

21. Adam Trexler, "Mediating Climate Change: Ecocriticism, Science Studies, and *The Hungry Tide*," in *The Oxford Handbook of Ecocriticism*, ed. Greg Garrard (Oxford: Oxford University Press, 2014), 205.

22. Trexler, "Mediating Climate Change," 205.

23. Ursula Kluwick, "Talking about Climate Change: The Ecological Crisis and Narrative Form," in *The Oxford Handbook of Ecocriticism*, ed. Greg Garrard (Oxford: Oxford University Press, 2014), 502.

24. Jericho Williams, "An Inscrutable Malice: The Silencing of Humanity in *The Ruins* and *The Happening*," in *Plant Horror: Approaches to the Monstrous Vegetal in Fiction and Film*, ed. Dawn Keetley and Angela Tenga (London: Palgrave Macmillan, 2016), 234.

25. Williams, "An Inscrutable Malice," 234.

26. E. Ann Kaplan, *Climate Trauma: Foreseeing the Future in Dystopian Film and Fiction* (New Brunswick, NJ: Rutgers University Press, 2016), 36.

27. Froula, "What Keeps Me Up at Night," 114.

28. Kaplan, *Climate Trauma*, 35.

29. Keetley, "Introduction," 6.

30. Keeker and Szabari qtd. in Keetley, "Introduction," 17.

31. Kaplan, *Climate Trauma*, 56.

32. Williams, "An Inscrutable Malice," 237.

33. Williams, "An Inscrutable Malice," 238.

34. Williams, "An Inscrutable Malice," 238.

BIBLIOGRAPHY

Briefel, Aviva, and Sam J. Miller. Introduction to *Horror after 9/11: World of Fear, Cinema of Terror*, 1–12. Edited by Aviva Briefel and Sam J. Miller. Austin: University of Texas Press, 2011.

Dixon, Wheeler Winston. "Introduction: Something Lost—Film after 9/11." In *Film and Television after 9/11*, edited by Wheeler Winston Dixon, 1–28. Carbondale: Southern Illinois University Press, 2004.

Froula, Anna. "What Keeps Me Up at Night: Media Studies Fifteen Years after 9/11." *Cinema Journal* 56, no. 1 (2016): 111–18. Accessed February 19, 2018. https://doi .org/10.1353/cj.2016.0056

Ghosh, Amitav. *The Great Derangement: Climate Change and the Unthinkable*. Chicago: University of Chicago Press, 2016.

Gordimer, Nadine. "Witness: The Inward Testimony." Government of India, Ministry of External Affairs, November 10, 2008. Accessed January 20, 2021, https://www.mea .gov.in/Speeches-Statements.htm?dtl/1754/WITNESS++THE+INWARD+TESTI.

Hantke, Steffen. "Academic Film Criticism, the Rhetoric of Crisis, and the Current State of American Horror Cinema: Thoughts on Canonicity and Academic Anxiety." *College Literature* 34, no. 4 (2007): 191–202. Accessed February 19, 2018. http://www.jstor.org/stable/25115464.

Kaplan, E. Ann. *Climate Trauma: Foreseeing the Future in Dystopian Film and Fiction*. New Brunswick: Rutgers University Press, 2016.

Keetley, Dawn. "Introduction: Six Theses on Plant Horror; or, Why Are Plants Horrifying?" In *Plant Horror: Approaches to the Monstrous Vegetal in Fiction and Film*, edited by Dawn Keetley and Angela Tenga, 1–30. London: Palgrave Macmillan, 2016.

Kluwick, Ursula. "Talking about Climate Change: The Ecological Crisis and Narrative Form." In *The Oxford Handbook of Ecocriticism*, edited by Greg Garrard, 502–16. Oxford: Oxford University Press, 2014.

Markert, John. *Post-9/11 Cinema: Through a Lens Darkly*. Lanham: Scarecrow Press, 2011.

McSweeney, Terence. *The "War on Terror" and American Film: 9/11 Frames per Second*. Edinburgh: Edinburgh University Press, 2014.

Morton, Timothy. *Dark Ecology: For a Logic of Future Coexistence*. New York: Columbia University Press, 2016.

Prorokova-Konrad, Tatiana. "The Ecohorror of *The Strain*: Plant Vampires and Climate Change as a Holocaust." In *The Global Vampire: Essays on the Undead in Popular Culture Around the World*, edited by Cait Coker, 189–99. Jefferson NC: McFarland, 2020.

The Happening. Directed by M. Night Shyamalan, performances by Mark Wahlberg, Zooey Deschanel, John Leguizamo, and Ashlyn Sanchez, 20th Century Fox, 2008.

Trexler, Adam. "Mediating Climate Change: Ecocriticism, Science Studies, and *The Hungry Tide*." In *The Oxford Handbook of Ecocriticism*, edited by Greg Garrard, 205–24. Oxford: Oxford University Press, 2014.

Williams, Jericho. "An Inscrutable Malice: The Silencing of Humanity in *The Ruins* and *The Happening*." In *Plant Horror: Approaches to the Monstrous Vegetal in Fiction and Film*, edited by Dawn Keetley and Angela Tenga, 227–41. London: Palgrave Macmillan, 2016.

Chapter Eight

The Global Impact of Fast Fashion

Understanding Sustainability and Social Justice Issues

Helen Liu and Alyssa Racco

OVERVIEW

This chapter addresses the risks and consequences that fast fashion poses and attempts to frame the complex story behind each article of clothing we purchase. How this story unfolds from raw materials to production processes to brand advertisement to discard practices has incredible relevance for climate-conscious consumers. The chapter also introduces sustainable fashion and explores the opportunities, approaches, and options that sustainable fashion provides to positively impact change in various aspects of our world, such as fundamental human rights and environmental protections. It aims to foster awareness about fast fashion and support educators in guiding students toward a better understanding of the factors that contribute to climate change. By participating in activities and reflective questioning about the fashion industry, students can have the opportunity to understand how detrimental consumer culture is because of capitalism-driven competition in relation to climate and social justice issues.

WHAT IS SUSTAINABILITY?

The concept of sustainability can be interpreted as the effort to engage with and utilize natural resources in less wasteful methods.[1] This term originated in 1987 as a policy concept in the Brundtland Report, which was conducted by the UN World Commission on Environment and Development with the aim to illuminate the environmental issues and dangers of scarce natural resources; it was also meant to inspire improved living conditions on a global scale.[2] Though this term can possess a broad scope that is multidimensional, it is typically understood to consist of three major aspects (social, economic, and environmental sustainability) that all aim to meet the current needs of the world without compromising the welfare of future generations.[3] These three foundational pillars of sustainability, also understood as people, profit, and planet, are interconnected with each other.[4] Every individual is impacted by actions, or inactions, in this interrelated world, with consequences intersecting across boundaries and generations.

Thus, sustainability is not just defined by its relationship with the environment, but also encompasses complex relationships with individuals, communities, and institutions.[5] Sustainability fosters a better understanding of the world and our place in it as we navigate through fluctuating landscapes to develop a consciousness and accountability for improving the social, economic, political, health, and ecological systems that support growth. It is not just limited to notions of preservation but entails change.[6] Sustainability encompasses equal harmonization of conservancy and creativity. Every individual not only needs to adapt to the evolving world but is required to practice such sustainable efforts within an applicable frame of time if there is hope to effectively address the destruction that is occurring from global warming at alarmingly rapid and devastating rates.[7]

Today's educators have the potential to shape the next generation of students to become conscious of their environmental responsibilities and become sustainable and responsible citizens of the world. Learning about sustainability allows students to develop a deep understanding of how humanity and the environment are intertwined. Teaching students to understand what fast fashion is and the systems that drive this industry can foster awareness and become a motivator for them to support sustainable fashion. Furthermore, it is important to acknowledge that this is not an effort to create blame or despair for students, but to educate them about these underlining systems and help them to reflect about their potential consumption in an effort to create agency, not participation, in this capitalist-driven society. It is important that educators advocate for corporate social responsibility as companies and brands play a crucial role in driving the agenda toward sustainability through the establishment of ethical codes of conduct, and the implementation of sustainable initiatives or strategies.[8] However, educators can also positively influence students by fostering consumer awareness, especially before students have established their own purchasing behavior, attitudes, and habits. Knowledge allows them an opportunity to deeply reflect on the systemic forces that drive consumer culture so they can make informed choices.

Introducing the consequences of fast fashion can help shape students' consumer behavior, which can have a positive impact on climate change. Patterns of consumption and interest can lead to the support or rejection of policies and initiatives.[9] With the spread of awareness regarding fast fashion, a surge of interest in sustainable and ethical fashion has emerged. This can potentially shape consumer values and behaviors that may ultimately lead to major changes that impact climate change. For example, in recent years veganism has been on the rise[10] as knowledge and awareness regarding animal cruelty have spread across the globe. This has led to major changes within the fashion industry,[11] as many corporations have adopted animal cruelty alternatives like faux leather or faux fur in order to meet the demands of consumers who want to adopt a fully cruelty-free lifestyle.[12]

The spread of consumer behavior and awareness have also led to other major changes toward the sustainable fashion initiative, including the success of ethical advocacy like PETA's cruelty-free campaigns. This campaign resulted in hundreds of major brands and designers, like Gucci, Chanel, and Burberry, to officially enact a ban against fur and solely sell animal-free clothing.[13] There have also been historic legislative changes occurring across the globe as a result of increased awareness regarding animal cruelty and ethical fashion. For instance, Israel became the first country in the world to completely ban the sale of fur for fashion, making their fashion market more environmentally friendly and encouraging other countries' governments to adopt similar regulations.[14]

Sustainability is a very popular concept in today's society, and might even be considered a trendy idea, as it is increasingly being promoted by individuals, various local and international organizations, and governments across the globe. Business establishments, producers, and consumers are beginning to preach the principles of sustainability in their policies and lifestyles.[15] Though these actions can definitely bring about positive change, as there can be growth in environmental awareness and opportunities to foster well-being and concern for the future, there needs to be meaningful intention.[16] Despite the positive attributes verbal promotions may have, there needs to be more action that goes beyond the surface level, with the core values and principles of sustainability promoted by individuals who truly understand the concept and are willing to actually put it into their practice, policies, and business processes.

It can be extremely easy for business corporations or brands to endorse their operations as sustainable, labelling themselves as "green" or "ethical" without actually embracing these values in any aspect of their procedures. These superficial, vague, and unsubstantiated promotions of sustainability ultimately deteriorate the fundamental values of the concept and practices. The "sustainability" that people think they are supporting may unfortunately lack effort or meaning. These contrived claims of embracing ethical or sustainable products or practices can be defined as "greenwashing," now utilized as a marketing tool to respond to consumer concerns, public relations, brand recognition, and positive corporation representation due to the trendiness of the concept, rather than a true passion for change within society.[17] Such non-concrete notions of sustainability can put the concept at risk, transforming it into a simple "feel good issue"[18] that lacks intention or stakes. There needs to be opportunities for individuals to learn and practice the fundamental values of sustainability, truly understanding the lasting impact their complex relationships and everyday actions have on the world. Recognizing and understanding the dark controversies or conflicts embedded within these issues is essential to evoke radical change within individual ways of living as well as in the operations of communities, businesses, and governments.

WHAT IS FAST FASHION?

The term *fast fashion* refers to clothing that is low-cost and often replicates the current luxury and popular fashion trends.[19] It is a hyperaccelerated business model that "involves increased numbers of new fashion collections every year, quick turnarounds and often lower prices. Reacting rapidly to offer new products to meet consumer demand is crucial to this business model."[20] In order to achieve this, fast fashion often "utilizes trend replication, rapid production, and low quality materials in order to bring inexpensive styles to the public."[21] The rise of major fast-fashion stores, such as H&M and Zara, has drastically changed the course of global fashion. Intense overproduction, decreased expenses, and the growth of more fashion seasons has led to accelerated rates of clothing production and trend turnover. These drastic changes have impacted modern consumers and their fashion purchasing behaviors and attitudes, potentially leading to individuals indulging in overconsumption, to perceive higher value returned for smaller investments, and ultimately to allow for the continued practice of unethical fashion production that serves the increasing demand for the industry.[22] Fast fashion succeeds based on the rapid fashion cycles, as the very nature of it involves a process that has steadfast return and incites disposability among consumers.

Fashion seasons that previously upheld a cycle of six months are now being promoted as new collections to the public within a few weeks before rapidly being transitioned out for the next collection.[23] With the growing quantities of fast-fashion stores emerging across the globe, these retailers can easily influence and engage with customers in faster and more drastic measures.[24] To maintain a steady flow of customers, fast-fashion retailers like Zara and H&M also regularly ensure that their stock is updated with the latest trends and popular items are being restocked.[25] Furthermore, with the technological advancements occurring alongside the fashion industry, it is no surprise that social media has become a tool utilized by retailers to successfully influence and promote their products to many consumers on a global scale. With the aid of social media, many retailers' fashion can be seen as familiar and relatable, especially when advertised by social media influencers or stylists that consumers follow and trust, becoming an essential part of many brands' marketing and endorsement strategies that possess a broader influence for potential customers.[26]

This fast-paced turnover in the industry has influenced individuals and the ways they consume fashion, with many developing the practice to consistently peruse through fashion retail websites or stores every few weeks in search of new trends or items. The limited volume of products and rapid shopping cycles also fosters a contradictory notion of mass exclusivity for consumers, which ultimately influences their impulsive shopping behavior in order to ensure that they do not miss out on any must-have trendy designs.[27] These notions of exclusivity also satisfy many consumer's need for developing their own uniqueness or individuality in order to enhance their self-representation and social image.[28] With retailers' consistent cycle of trendy and "exclusive" fashion products, many fast-fashion brands are quickly gaining brand recognition and a following of loyal consumers.[29] This allows brands to solidify their position in the industry and their perceived value allows them to "command a price premium in the marketplace."[30] This brand consciousness, alongside the available dispensable income consumers possess, results in fast-fashion retailers exploiting individuals and the instant satisfaction many experience when they engage in impulsive shopping practices and behaviors.[31]

SUSTAINABLE FASHION

Despite the immense culture of fast fashion, there is also an emerging movement of anti-consumerism, with many individuals becoming aware of the realities involved in the fashion industry and upholding an anti-market attitude, such as boycotting fast-fashion brands or resisting mindless overconsumption.[32] The earliest anti-fur action appeared in the 1980s, and with revelation of various sweatshop scandals in the late 1990s, there was increasing awareness and pressure from society to implement more ethical and effective policies and operations for retail factories.[33] Individuals are also becoming conscious of their role in the industry, and how their distinct consumption habits can further the production of businesses and demand, producing "an ongoing cycle of appetite, simultaneously voracious and insatiable."[34] Social media platforms have become one of the most common methods for individuals, in particular young people, to learn and access information.[35] Despite the potential for social media platforms to contribute to fast-fashion consumption, it has also played a major role in engaging and increasing audiences understanding of ethical fashion consumption. Through social media, many individuals have started to become aware of their own product consumption, which is reflective in the increase of media representation these platforms have for the environment and sustainability.[36] In particular, popular platforms like Instagram, TikTok, and YouTube allow individuals to interact with influencers that they trust and connect with. Individuals are able to embrace fashion brand and chain awareness and to learn about other sustainable fashion activities they can participate in, such as thrifting, that allows them to still participate in fashionable trends but in more ethical and sustainable ways.

Fostering this awareness and exposing the unethical and devastating realities involved in fast fashion can inform anti-consumerism attitudes and practices that will combat this ongoing cycle of demand and production. Anti-consumerism is dependent on the everyday practices and choices of individuals. These simple choices cultivate deep critical discussions and reflections regarding the market and industry, serving to "anchor subjectivities in constructed and heavily mediated narratives of lifestyles, self-hood, community, and identity."[37] Through this, individuals can develop a sense of responsibility and begin to understand the significance of their actions and the short and long-term consequences involved. Such knowledge and understanding can lead to an evolution and reinvention of their identities and realign their paths to more environmentally conscious and ethical fashion.[38]

Thus, the sustainable fashion movement, also known as eco, green, or ethical fashion, has been on the rise in the past decade and is becoming increasingly mainstream.[39] Sustainable fashion is included in the slow fashion movement, which is founded on a model that encompasses sustainability values, such as ethical and fair labor conditions, and an effort to decrease environmental devastation across the globe.[40] The core principles of sustainable fashion also include the process of working toward maintainability, with the objective to "create a system which is supportable indefinitely in terms of human impact on the environment and social responsibility."[41] Though sustainable fashion can be a broad term that is associated with a variety of ideals or practices, such as ethical business operations, shopping used or second-hand, organic and ecologically friendly resources, and certain certifications, the movement ultimately challenges the fast-fashion industry by disintegrating the existing narratives and restrictions between corporations and stakeholders, slowing fashion production, and shifting notions of self-concept to focusing on collective consciousness that considers the well-being of future generations and fosters opportunities for choices that empower transformation.[42]

With this increasing consumer awareness of the environmental and social concerns involved with fashion,[43] many fast-fashion retailers and corporations are attempting to adapt their brand to more sustainable and ethically conscious operations to appeal toward consumers' changing behavior and shifts in the fashion industry. Due to increased awareness of and interest in sustainability, many fast-fashion brands have faced immense criticism due to unethical conditions or practices within their production processes or textile choices. As consumers begin to question where the clothing is from and the ethical conditions of these brands, many corporations are quickly learning about the social responsibilities they have. In turn, they are modifying their business models to adopt sustainable fashion that meets the demands and values of consumers, in hopes of retaining positive success and brand reputation.[44] For example, when Nike was exposed for having inhumane production practices through their use of sweatshops, the company prioritized efforts toward changing their supply chains and adopting more ethical labor conditions in the last two decades.[45] Companies make similar attempts by introducing a new product extension or initiative through an existing brand that individuals may already have prevailing knowledge of or recognize, reducing the perceived risk for both corporations and consumers. These sustainable or ethical fast-fashion strategies and actions have already been implemented with a variety of fast-fashion retailers, such as H&M Conscious, Zara Join Life, Top Shop Considered, and UNIQLO Sustainability, all with mission statements emphasizing corporate social responsibility and the objective to create fashion through sustainable and ethical practices that take care and consideration of humans and the environment.[46]

HOW DO YOU KNOW A BRAND IS SUSTAINABLE?

Despite these sustainable changes from fast-fashion brands, it is still extremely easy for corporations to promote and claim ethical practices in their operations in order to appeal to consumers without genuinely delivering on these values and actions. It can be confusing for many consumers to understand whether a brand is truly sustainable or ethical without proper awareness of the potential tactics companies implement in their operations, in addition to the proper recognition of certifications or research conducted when trying to shop sustainably. In order to understand some of the ethical purposes and initiatives behind brands, it is first important to look at the major sustainable development goals (SDGs) that were established worldwide. In 2000, the United Nations (UN) created the "UN Global Compact," which outlined the world's leading sustainability and developmental initiatives for corporations. There were seventeen goals outlined that aimed to "unite all global stake-holders to end extreme poverty, fight inequality and injustice, and protect our planet."[47] These seventeen goals included

- No Poverty
- Zero Hunger
- Good Health and Well-Being
- Quality Education
- Gender Equality
- Clean Water and Sanitation
- Affordable and Clean Energy
- Decent Work and Economic Growth
- Industry Innovation and Infrastructure
- Reduced Inequalities
- Sustainable Cities and Communities
- Responsible Consumption and Production
- Climate Action
- Life below Water
- Life on Land
- Peace, Justice, and Strong Institutions
- Partnerships for the Goals

Understanding these sustainable goals will help guide consumers to reflect and do research into the brands they are buying as well as foster an under-standing of the brand's purpose and longstanding branding strategies and goals. Understanding a brand's exact ethical initiative will allow individuals to consider whether a brand is genuine in their mission toward sustainability, whether ethical values are truly being respected and delivered in their model.

Consumers can also recognize and review sustainability certifications as supporting evidence to understand the environmental and human impact of companies. Sustainability certifications and standards were designed to provide consumers identifiable information regarding a product and whether it was environmentally friendly or independently qualified as sustainable.[48] These certifications can assist consumers in understanding whether a brand's vague claims of "fair-trade" or "cruelty-free" can be substantiated with legitimate qualifications accredited by third parties. These certifications are not limited to, but can include:

- Better Cotton Initiative—A cotton sustainability program aims to improve the conditions of cotton production worldwide to foster positive environmental and social changes.[49]
- Cradle to Cradle Certified—Determines whether a product has been made either entirely recyclable or biodegradable and developed under the lowest and most sustainable manufacturing processes.[50]
- Fairtrade—Indicates that a product has achieved a level of social, economic, and environmental standards that supports sustainable development and producers in developing nations to achieve equal trade relations.[51]
- Global Organic Textile Standard (GOTS)—Signifies that a product is organic throughout every aspect of production, encompassing biodegradable products and low impact waste management and fair practices.[52]
- Fair Wear Foundation—A nonprofit organization with the objective to partner with brands, companies, factories, and trade unions to validate and increase the workplace standards for apparel laborers in eleven nations in Asia, Africa, and Europe.[53]
- Ethical Trading Initiative (EI)—Aims to protect the rights of workers in supply chains worldwide, ensuring that employment was freely chosen, no child labor is used, and that conditions are safe and fair.[54]
- Organic Content Standard (OCS)—Verifies that a product is made from organic material or organically grown throughout the whole production process.[55]
- People for the Ethical Treatment of Animals (PETA)—An animal rights organization with the goal to reduce animal suffering in all industries or aspects in the world, such as the fashion industry or animal testing in laboratories.[56]
- Sustainable Apparel Coalition—Certifies that any type of garment whether it be apparel, textiles, or footwear has been produced in sustainable conditions that does not create excessive environmental destruction and has a positive social impact.[57]

More certifications can be found at Apparel Entrepreneurship's *Sustainability Certification Guide*.[58]

These certifications allow consumers to gain knowledge and awareness of what the standards are for a sustainable brand, fostering responsible commerce and the ability for consumers to start critically questioning and reflecting about the brands they are shopping. For example, understanding these certifications can allow consumers to determine whether a brand is ethical or sustainable by questioning:

- Is the brand contributing to environmental or social destruction in any way?
- Are the products made with organic materials that are biodegradable or recyclable?
- Does the manufacturing process of the brand have a low environmental impact?
- Is the brand protecting the rights of their workers? Are work conditions fair and safe?
- Is the brand engaging with any exploitive child labor?
- Does the brand engage in animal cruelty?
- Does the brand have any positive objectives to improve conditions in the fashion industry?
- Is the brand an independent or local business?
- Where is the brand made?
- Does the brand have any ethical certifications or qualifications?

Many consumers also need to consider the potential duplicitous tactics corporations may utilize in order to appeal to the public and meet the technicalities of their sustainable claims, as many brands may appear to embody ethical or sustainable values when first considered. For example, a brand may assert that they are fair trade, ethical, or sustainable, but in reality, conducting research on their website may reveal only a limited or small collection of these products among fast-fashion items. Furthermore, brands can maneuver around claims that items are "Made" and produced in developed countries, such as the United States or Canada, by having the items sewn or stitched there. However, all raw materials and various resources may have come from sweatshops or other unethical sources abroad in developing nations. Many brands are also contradictory in their actions, as some may appear to uphold ethical values and social responsibilities by donating a portion of their profits to charities or causes; however, many times the items they are producing have been conducted in unfair and harmful conditions with huge negative environmental impacts.[59]

Learning about these sustainable development goals and certification standards ensures that consumers can go beyond superficial evaluations of fashion brands to understand the full story. Students can learn about this through educational opportunities, such as engaging in a research group project like a brand investigation. Educators can choose to conduct an informal poll of the most common brands preferred by the class (e.g., Nike, Toms, Zara, H&M), and assign each group a different brand to investigate. Utilizing the list of reflective questions provided, students can answer each inquiry with the concluding goal to determine whether or not their brand is ethical. Students are encouraged to collaborate and to utilize a variety of research methods during their brand investigation. This could include conducting a thorough examination of the brand or company website to understand their ethical position, identifying what certifications and qualifications the brand holds, and determining if any "sneaky tactics" were utilized to falsely attract consumers.

Through this project, students will be able to develop consciousness around their consumption of ethical products and understand the impact brands have on the environment. Each group can share these findings through a final group presentation to their peers, showcasing their research and evidence to the rest of the class. This allows each student to understand how they may be contributing to climate and social injustice through their support of these common and popular brands or companies. Furthermore, if students determine that their brand is unethical, they may also be encouraged to suggest what potential changes are available for the brand to adopt in order to help make them more ethical. This is especially pertinent for young individuals who may not have established their own values and attitudes toward fashion and their product consumption. Students need the opportunity to learn about sustainable fashion, as individuals who are not passionate about this topic or remain unaware are unlikely to scrutinize the brands they are consuming. Thus, education is empowering for individuals to understand their role as consumers in the fashion industry, potentially influencing their purchasing practices, values, and attitudes. In doing so, students are able to cultivate deeper understandings and values of sustainability, including how their investments in sustainable fashion have worldwide impacts today and affect the welfare of generations to come.[60]

HUMAN RIGHTS AND FAST FASHION

What Are Human Rights?

We will look to the United Nations for the universal definition on human rights because it is a long-standing international organization to which 193 independent states belong.[61] The four foundational purposes and principles of this organization are outlined in a charter that all members must follow. The initial Charter was signed on June 26, 1945, by fifty countries at the "United Nations Conference on International Organization,"[62] but was not finalized until October 24 of that same year. The United Nations continue to meet and readjust its focus as the world evolves and new needs present themselves. Important to note, however, is the World Conference on Human Rights, held in Vienna, Austria, in 1993. Here, "contending interpretations of the scope of humans rights"[63] were ratified through clarification that the "universal nature of these rights and freedoms is beyond question."[64] Today, the overarching purpose of the Charter is to "maintain international peace and security."[65] A gigantic endeavor in itself, the following three principles work to fulfill the first: to cultivate friendly relations between nations; to solve international problems and promote human rights; and to harmonize "the actions of nations in the attainment of these common ends."[66]

As a result of the Charter and the World Conference on Human Rights, the United Nations defines human rights as

> rights we have simply because we exist as human beings—they are not granted by any state. These universal rights are inherent to us all, regardless of nationality, sex, national or ethnic origin, color, religion, language, or any other status. They range from the most fundamental—the right to life—to those that make life worth living, such as the rights to food, education, work, health, and liberty.[67]

Without these basic human rights, life as we know it would be vastly different. These basic rights and freedoms are embedded within every action we take, though we may not even be aware of it.

Classroom Activity Idea

Prompt: How do human rights impact our daily life?

It is undeniable that the rights and freedoms granted to us are used on a daily basis. We are accessing them right now through education! Educators may use the question "How do human rights impact your daily life?" as a prompt to foster discussion an explore the topic. At first, have students reflect on their own lives and then read Mehreen Tariq Ghani's article "Inside the Ugliness of the Fast Fashion Industry."[68] Have students contrast the ways they imagine human rights impact the children mentioned in the article to their own experiences. The article raises hard hitting questions about how brands like Zara, H&M, and Forever 21 are able to produce new clothing so often and keep their prices so low by explaining that the mainly women and children in poor countries are paid "starvation wages for working in horrific conditions."[69] Without being too graphic, and therefore classroom-appropriate, the article provides statistics that vividly paint a picture of what it is like to be a child laborer for fast-fashion conglomerates. Using stories, especially news articles, to engage students in child labor concerns, environmentalism, and climate-related issues allows them to make meaningful connections with the individuals centered in the narratives because they are real. When students reflect that "this could be me," they are able to emphasize with the situation of these children, who may be about the same age as them, and feel inspired to help make a difference. Examples of experiences to discuss include, but are not limited to, education, food and water, proper clothing, home conditions, and safety concerns.

CHILD LABOR

What Is Child Labor?

The International Labor Organization (ILO), a specialized agency of the United Nations, defines child labor as "work that deprives children of their childhood, their potential and their dignity, and that is harmful to physical and mental development."[70] The definitions extend to work that is "mentally, physically, socially, or morally dangerous and harmful for children; and/ or interferes with their schooling by: depriving them of the opportunity to attend school; obliging them to leave school prematurely; or requiring them to attempt to combine school attendance with excessively long and heavy work."[71] The ILO acknowledges that child labor may take many different forms, but prioritizes the elimination of the worst forms which include all forms of slavery, trafficking of children, child prostitution, using children for the production of pornography, using a child for illicit activities such as, but not limited to, drug trafficking, and any work that is harmful to the child's health, safety, or moral development.[72]

Child Labor and Human Rights

The ILO and UN exist to protect the rights of all children and hope to eradicate child labor by 2025. Both organizations "emphasize that freedom from child labour is a human right and that the elimination of child labour is a universal and fundamental value."[73] The following case study can be used in class to give historical context or perhaps give way to a new conversation about the importance of children in society and how they in themselves, rather than as a commodity for gain, deserve a full life.

Special Case: Mary Ellen Wilson

Mary Ellen Wilson, born in March of 1864 in New York, is seen as the catalyst for understanding the need to protect children and guarantee them their own rights. Not long after Mary Ellen's birth, her father died, leaving her mother Francis to raise her alone. In order to work, Francis sent Mary Ellen to live with another woman named Martha Score whom she would pay to look after her while she worked. Soon, Francis was unable to keep up with payments, so Martha lied and told her Mary Ellen had died, when in reality, she had surrendered her to the Department of Charities.[74]

While a ward of the state, Thomas and Mary McCormack claimed that they were Mary Ellen's biological parents and, without proof, they were able to take Mary Ellen home with them. After Thomas passed away, Mary McCormack remarried and relocated. It was only after this relocation that neighbors noticed Mary Ellen's condition: "it was no secret to anyone that the little girl's foster mother constantly beat, burned, whipped and cut her, locked her in a closet for hours on end, forced her to do heavy labour and to sleep on the floor and forbade her from ever going outside during the day."[75] As a result, a concerned neighbour, Etta Wheeler, took it upon herself to gain access into Mary Ellen's home, under the false pretense of needing assistance, in order to assess the situation for herself. Wheeler sought help from the authorities but was told that "the child should be thankful she was not on the streets."[76]

Unsatisfied with the response from authorities, Wheeler took it upon herself to reach out to Henry Bergh, the leader and founder of the American Society for the Prevention of Cruelty to Animals (ASPCA). Bergh supported Wheeler's outrage and had a lawyer from the ASPCA work Mary Ellen's case. The lawyer argued "that laws protecting animals from abuse should not be greater than law protecting children."[77] Wheeler and Bergh were successful and substantiated the idea that "children are *people*—not commodities to be owned nor servants to be exploited, and certainly not punching bags to take one's anger out on."[78]

Child Labor in Our World Today

According to the most recent Global Estimates of Child Labor report by International Labour Office, a total of "152 million children—64 million girls and 88 million boys—are in child labour globally, accounting for almost one in ten of all children worldwide."[79] What may come as a surprise is that this issue is not solely a "low-income country problem"; it affects children in even the wealthiest of countries. The report asserts: "84 million children in child labour, accounting for 56 per cent of all those in child labor, actually live in middle-income countries, and an additional 2 million live in high-income countries."[80] By the report's definition, a middle-income country and a gross national income per capita ranging from US$1,046– $12,735 in comparison to US$12,736 or more, is a high-income country.[81] Moreover, approximately seventy-three million of these children work in conditions so hazardous it directly "endangers their health, safety and moral development."[82]

Child Labor and Fast Fashion

Child labor is a part and parcel of the fast-fashion industry. Cheap and unskilled labor,[83] easily manageable obedient workers, and being an easy target[84] are just some of the reasons child labor remains a particular issue for fashion. Unfortunately, child labor is extremely difficult to prove and even discover: "employers get away with it because the fashion supply chain is hugely complex and it is hard for companies to control every stage of production. That makes it possible to employ children without big brands and consumers ever finding out."[85]

Cases

Utilizing SOMO's fact sheet, case studies can also be an opportunity for students to analyze child labor in the fashion industry. The first potential case that can be used speaks to the role young girls are most often required to do: cotton seed production. This task, a requirement for textiles to be mass manufactured, is dangerous and is still ongoing today.[86] The second case addresses a scheme that girls in India can fall trap to: girls are hired for multiyear contracts to work at textile spinning mills in order to be able to pay for their dowry in the future.[87] The final case, centered on children in Uzbekistan, addresses the policies in place that force children, as young as nine, to leave school and help with the cotton harvest.[88] These cases can be used as an activity in classes to highlight the different tasks that child laborers in the fast-fashion industry are doing. Alternatively, each mini case study can be used to imagine a character sketch of a child their own age.

THE ENVIRONMENT AND FAST FASHION

The Impact of Fast Fashion

Clothing is not only a well-practiced social norm, but a necessity to protect our bodies from harm. However, what we choose to wear "has an embedded environmental cost in terms of energy, water, land and chemicals used."[89]

Energy

The production of textiles produces approximately "1.2 billion tonnes of CO_2 [per] year—more than international flights and maritime shipping combined."[90]

Water

Exorbitant amounts of water are used to produce clothing, making the fashion industry "the second largest consumer industry of water."[91] To put this into perspective, roughly 700 gallons of water are used to produce a single shirt made of cotton.[92] Now consider the rates at which fast-fashion brands are producing these garments—the amount of water these brands consume is significantly more than other fashion brands. Consider Zara: "Zara changes its clothing designs every two weeks on average, while competitors change their designs every two to three months. It carries about 11,000 distinct items per year in thousands of stores worldwide compared to competitors that carry 2,000 to 4,000 items per year."[93]

Land

Clothing made from natural fibers requires the use of land to grow said fibers (think cotton). The mass production of these natural fibers can impact biodiversity within ecosystems and promote deforestation so that land can be repurposed: "The fashion industry is projected to use 35% more land for fibre production by 2030—an extra 115 million hectares that could be used to grow crops for an increasing population or preserve forest to store carbon."[94] Most clothing from fast-fashion brands, however, are made from synthetic fibers which come from "virgin plastics." Although these fibers "have less impact on water and land than cotton, [they] emit more greenhouse gases per kilogram,"[95] and take "hundreds of years to biodegrade."[96]

What Can We Do?

First and foremost, we must educate ourselves. The "Transparency Index" is an annual report of 250 brands "ranked according to how much they disclose about their social and environmental policies, practices and impacts."[97] The charts in this report give a clear idea about which brands are moving toward maintaining safe working environments for all workers, paying fair wages, and ecofriendly production practices. At the same time, it indicates which brands refuse to disclose information; as a consumer, that raises the question: why is it a secret?

Unfortunately, knowing is not nearly enough. As consumers we must act on this knowledge and buy responsibly. Investing in pieces that will last for years rather than uber-trendy pieces that last for a season is a great place to begin. Thinking about recycling will not make the cut either; the millions of tons of clothing that are discarded annually end up either in a "landfill or [burnt]," and less than 1% of fibres used to make garments are recycled into new clothing."[98] This goes back to the idea that the quality of clothing and the fibres being used is a key consideration when purchasing clothing. Sustainable fibers are always a better option because they last longer. With that said, a very sustainable option to consider when shopping is thrifting. This uber-trendy activity—popularized by YouTube and TikTok creators—is a great way to repurpose and reuse clothing that already exists. This in turn eliminates the need for new fibers to be grown and new clothing to be created. Nonetheless, we must consume less and keep the pieces for longer if we intend to keep our planet alive and healthy.

Activity Idea

Within the class, students can set up their own "Free Market." This idea would require students to bring in items of clothing they had planned to discard or even donate. Clothing items would be displayed, and students could "shop" for pieces. This could even be extended to be a schoolwide project. Keeping items free makes the market accessible to all students and therefore makes the activity inclusive. In addition, it teaches students the value in thrifting and instills habits that will ideally be acted upon outside of school.

CONCLUSION

Beyond the individual consumer, it is imperative that we demand political action in the form of industry regulation. Without change from the top, the irreversible ramifications of the fast-fashion industry will continue to plague our planet. Banning together and protesting so that a collective voice is heard could push government reforms and political intervention in a way that will consider the future of planet before the profit line. This kind of widescale mandate is what we ultimately need in order to eliminate fast fashion and preserve our planet.

NOTES

1. Leslie Thiele, *Sustainability* (Cambridge: John Wiley & Sons, 2013), 3.
2. Tom Kuhlman and John Farrington, "What Is Sustainability?," *Sustainability* 2, no. 11 (2010): 3437, https://doi.org/10.3390/su2113436
3. Kuhlman and Farrington, "What Is Sustainability?," 3438.
4. Thiele, *Sustainability*, 5.
5. Joy Annamma, John R. Sherry, Alladi Venkatesh, Jeff Wang, and Ricky Chan, "Fast Fashion, Sustainability, and the Ethical Appeal of Luxury Brands," *Fashion Theory* 16, no. 3 (2012): 274, https://doi.org/10.2752/175174112X13340749707123

6. Thiele, *Sustainability*, 3.

7. Thiele, *Sustainability*, 4.

8. Aelsea Shephard and Sanjukta Pookulangara, *The Slow Fashion Process: Rethinking Strategy for Fast Fashion Retailers* (Florida: CRC Press, 2013), 11.

9. Christina Tobler, Vivianne H. M. Visschers, and Michael Siegrist, "Addressing Climate Change: Determinants of Consumers' Willingness to Act and to Support Policy Measures," *Journal of Environmental Psychology* 32, no. 3 (2012): 197, https://doi.org/10.1016/j.jenvp.2012.02.001

10. Outi Lundahl, "Celebrities and the Media in the Rise of Veganism," *Consumption Markets & Culture* 23, no. 3 (2020): 243.

11. Yeong-Hyeon Choi and Kyu-Hye Lee, "Ethical Consumers' Awareness of Vegan Materials: Fake Fur and Fake Leather," *Sustainability* 13, no. 1 (2021): 2.

12. London Design Collective, "Why Vegan Fashion Has Grown in Popularity," London Design Collective, accessed August 17, 2021, https://londondesigncollective.com/fashion/why-vegan-fashion-has-grown-in-popularity/.

13. PETA, "PETA Pushes Fashion Brands to Move Beyond Using Fur—and Wins," PETA, accessed August 17, 2021, https://www.peta.org/features/fur-free-companies-brands-that-banned-fur/.

14. Liam Giliver, "Historic: Israel Becomes World's First Country to Ban Fur Sales," Plant Based News, accessed August 17, 2021, https://plantbasednews.org/culture/ethics/israel-becomes-worlds-first-country-to-ban-fur-sales/.

15. Thiele, *Sustainability*, 6.

16. Thiele, *Sustainability*, 8.

17. Thiele, *Sustainability*, 6.

18. Thiele, *Sustainability*, 8.

19. Annamma, Sherry, Venkatesh, Wang, and Chan, "Fast Fashion," 275.

20. House of Commons, *Fixing Fashion: Closing Consumption and Sustainability*, Environmental Audit Committee, 2019, publications.parliament.uk/pa/cm201719/cmselect/cmenvaud/1952/1952.pdf.

21. Audrey Stanton, "What Is Fast Fashion, Anyway?," *The Good Trade*, accessed February 8, 2021, www.thegoodtrade.com/features/what-is-fast-fashion.

22. Tara Stringer, Gary Mortimer, and Alice Ruth Payne, "Do Ethical Concerns and Personal Values Influence the Purchase Intention of Fast-Fashion Clothing?," *Journal of Fashion Marketing an Management: An International Journal* 24, no. 1 (2019): 99.

23. Annamma, Sherry, Venkatesh, Wang, and Chan, "Fast Fashion," 275.

24. Nebahat Tokatli, "Global Sourcing: Insights from the Global Clothing Industry—the Case of Zara, A Fast Fashion Retailer," *Journal of Economic Geography* 8, no. 1 (2008): 23, https://doi.org/10.1093/jeg/lbm035

25. Annamma, Sherry, Venkatesh, Wang, and Chan, "Fast Fashion," 275.

26. Nawaz Ahmad, Ashiq Rubab, and Atif Salman, "The Impact of Social Media on Fashion Industry: Empirical Investigation from Karachiites," *Journal of Resources Development and Management* 7 (2015): 1.

27. Annamma, Sherry, Venkatesh, Wang, and Chan, "Fast Fashion," 276; Tokatli, "Global Sourcing," 23.

28. Jin Su and Aihwa Chang, "Factors Affecting College Students' Brand Loyalty toward Fast Fashion," *International Journal of Retail & Distribution Management* 46, no. 1 (2018): 94.

29. Su and Chang, "Brand Loyalty," 93.

30. Su and Chang, "Brand Loyalty," 96.

31. Annamma, Sherry, Venkatesh, Wang, and Chan, "Fast Fashion," 276.

32. Annamma, Sherry, Venkatesh, Wang, and Chan, "Fast Fashion," 277.

33. Louise Lundblad and Iain A. Davies, "The Values and Motivations behind Sustainable Fashion Consumption," *Journal of Consumer Behavior* 15, no. 2 (2016): 149, https://doi.org/10.1002/cb.1559

34. Annamma, Sherry, Venkatesh, Wang, and Chan, "Fast Fashion," 277.

35. Jin Mao, "Social Media for Learning: A Mixed Methods Study on High School Students' Technology Affordances and Perspectives," *Computers in Human Behavior* 33 (2014): 221 https://doi.org/10.1016/j.chb.2014.01.002.

36. Stephanie Anne Nicole Bedard and Carrie Reisdorf Tolmie, "Millennials' Green Consumption Behaviour: Exploring the Role of Social Media," *Corporate Social Responsibility and Environmental Management* 25, no. 6 (2018): 1388, https://doi.org/10.1002/csr.1654

37. Sam Binkley, "Liquid Consumption: Anti-Consumerism and the Fetishized De-fetishization of Commodities," *Cultural Studies* 22, no. 5 (2008): 601.

38. Annamma, Sherry, Venkatesh, Wang, and Chan, "Fast Fashion," 277.

39. Arch G. Woodside and Monica B. Fine, "Sustainable Fashion Themes in Luxury Brand Storytelling: The Sustainability Fashion Research Grid," *Journal of Global Fashion Marketing* 10, no. 2 (2019): 113, https://doi.org/10.1080/20932685 .2019.1573699

40. Claudia E. Henninger, Panayiota J. Alevizou, and Caroline J. Oates, "What Is Sustainable Fashion?," *Journal of Fashion Marketing and Management: An International Journal* 20, no. 4 (2016): 400, https://doi.org/10.1108/JFMM-07-2015-0052

41. Arch G. Woodside and Monica B. Fine, "Sustainable Fashion Themes in Luxury Brand Storytelling: The Sustainability Fashion Research Grid," *Journal of Global Fashion Marketing* 10, no. 2 (2019): 113, https://doi.org/10.1080/20932685 .2019.1573699

42. Henninger, Alevizou, and Oates, "What Is Sustainable Fashion?," 401.

43. Aelsea Shephard and Sanjukta Pookulangara, *The Slow Fashion Process*, 11.

44. Aelsea Shephard and Sanjukta Pookulangara, *The Slow Fashion Process*, 15.

45. Woodside and Fine, "Sustainable Fashion Themes," 120.

46. Stringer, Mortimer, and Payne, "Ethical Concerns," 100.

47. United Nations, "17 Goals to Transform Our World," United Nations Global Impact, accessed February 8, 2021, https://www.unglobalcompact.org/sdgs/17-global -goals.

48. Susan M. Harris, "Green Tick™: An Example of Sustainability Certification of Goods and Services," *Management of Environmental Quality: An International Journal* 18, no. 2 (2007): 171, https://doi.org/10.1108/14777830710725830

49. Better Cotton Initiative, "About Better Cotton," accessed February 8, 2021, http://bettercotton.org.

50. Cradle to Cradle Certified, "What Is Cradle to Cradle Certified?," accessed February 8, 2021, http://www.c2ccertified.org.

51. Fairtrade, "What Is Fairtrade?," accessed February 8, 2021, http://www.fairtrade.net.

52. Global Organic Textile Standard (GOTS), "The Standard," accessed February 8, 2021, https://global-standard.org/the-standard.

53. Fair Wear Foundation, "About Us," accessed February 8, 2021, https://www.fairwear.org.

54. Ethical Trading Initiative (EI), "About ETI," accessed February 8, 2021, https://www.ethicaltrade.org.

55. Organic Content Standard (OCS), "What Is the OCS?," accessed February 8, 2021, http://textileexchange.org.

56. People for the Ethical Treatment of Animals (PETA), "About PETA," accessed February 8, 2021, https://www.peta.org.

57. Sustainable Apparel Coalition, "The SAC," accessed February 8, 2021, https://apparelcoalition.org.

58. Apparel Entrepreneurship's Sustainability Certification Guide (2021), accessed February 14, 2021, https://www.apparelentrepreneurship.com/your-guide-to-sustainability/.

59. Ellie Hughes, "How to Tell If a Brand Is Ethical or Not," *The Practivist*, accessed February 8, 2021, http://thepractivist.com/blog/how-to-tell-if-a-brand-is-ethical-or-not.

60. Thiele, *Sustainability*, 14.

61. United Nations, "What Are Human Rights," OHCHR, accessed February 8, 2021, https://www.ohchr.org/en/issues/pages/whatarehumanrights.aspx.

62. United Nations, "History of the United Nations," United Nations: Peace, Dignity and Equality on a Healthy Planet, accessed February 8, 2021, www.un.org/en/sections/history/history-united-nations/index.html.

63. Population Council, "Vienna Declaration on Human Rights," *Population and Development Review* 19, no. 4 (1993): 877, https://doi.org/10.2307/2938429

64. United Nations Human Rights, "Vienna Declaration and Programme of Action," OHCHR, accessed February 8, 2021, www.ohchr.org/en/professionalinterest/pages/vienna.aspx.

65. United Nations, "Chapter 1," United Nations: Peace, Dignity and Equality on a Healthy Planet, accessed February 8, 2021, www.un.org/en/sections/un-charter/chapter-i/index.html.

66. United Nations, "Chapter 1."

67. United Nations Human Rights, "What Are Human Rights?," OHCHR, accessed February 8, 2021, https://www.ohchr.org/en/issues/pages/whatarehumanrights.aspx.

68. Mehreen Tariq Ghani, "Inside the Ugliness of the Fast Fashion Industry," *The Medium*, July 9, 2020, https://medium.com/maverickyouth/inside-the-ugliness-of-the-fast-fashion-industry-ac40f6a24e01.

69. Ghani, "Fast Fashion Industry."

70. International Labour Organization, "What Is Child Labour?," IPEC, accessed February 8, 2021, www.ilo.org/ipec/facts/lang--en/index.htm.

71. International Labour Organization, "What Is Child Labour?"

72. International Labour Organization, "What Is Child Labour?"

73. International Labour Office, "Global Estimates of Child Labour: Results and Trends, 2012–2016," (2017): 19, www.ilo.org/wcmsp5/groups/public/@dgreports/@dcomm/documents/publication/wcms_575499.pdf.

74. Giulia Montanari, "Mary Ellen Wilson: When Abused Children Had Fewer Rights Than Pets," *Medium*, August 12, 2020, medium.com/history-of-yesterday/mary-ellen-wilson-the-abused-child-rescued-by-animal-rights-activists-9dfb5b1f100a.

75. Montanari, "Mary Ellen Wilson."

76. Montanari, "Mary Ellen Wilson."

77. Montanari, "Mary Ellen Wilson."

78. Montanari, "Mary Ellen Wilson."

79. International Labour Office, "Global Estimates," 11.

80. International Labour Office, "Global Estimates," 13.

81. International Labour Office, "Global Estimates," 32.

82. International Labour Office, "Global Estimates," 11.

83. SOMO, "Fact Sheet Child Labour in the Textile & Garment Industry," (2014): 3, www.somo.nl/wp-content/uploads/2014/03/Fact-Sheet-child-labour-Focus-on-the-role-of-buying-companies.pdf.

84. Josephine Moulds, "Child Labour in the Fashion Supply Chain," *The Guardian and Media*, accessed February 8, 2021, labs.theguardian.com/unicef-child-labour/.

85. Moulds, "Child Labour."

86. SOMO, "Fact Sheet Child Labour," 2.

87. SOMO, "Fact Sheet Child Labour," 4.

88. SOMO, "Fact Sheet Child Labour," 3.

89. House of Commons, "Fixing Fashion: Closing Consumption and Sustainability," *Environmental Audit Committee*, (2019): 28, https://publications.parliament.uk/pa/cm201719/cmselect/cmenvaud/1952/1952.pdf.

90. House of Commons, "Fixing Fashion," 28.

91. Rashimila Maiti, "Fast Fashion: Its Detrimental Effect on the Environment," Earth.org, January 29, 2020, earth.org/fast-fashions-detrimental-effect-on-the-environment/.

92. Maiti, "Fast Fashion."

93. Mhugos, "Zara Clothing Company Supply Chain," SCM Globe, January 4, 2020, www.scmglobe.com/zara-clothing-company-supply-chain.

94. House of Commons, "Fixing Fashion," 30.

95. House of Commons, "Fixing Fashion," 31.

96. Maiti, "Fast Fashion."

97. Fashion Revolution, "Fashion Transparency Index 2020," *Issuu* (2020): 1, https://issuu.com/fashionrevolution/docs/fr_fashiontransparencyindex2020.

98. Angeli Mehta, "Beyond Recycling: Putting the Brakes on Fast Fashion," *Reuters Events*, April 28, 2019, www.reutersevents.com/sustainability/beyond-recycling-putting-brakes-fast-fashion.

BIBLIOGRAPHY

Ahmad, Nawaz, Rubab Ashiq, and Atif Salman. "The Impact of Social Media on Fashion Industry: Empirical Investigation from Karachiites." *Journal of Resources Development and Management* 7 (2015): 1–8.

Annamma, Joy, John R. Sherry, Alladi Venkatesh, Jeff Wang, and Ricky Chan. "Fast Fashion, Sustainability, and the Ethical Appeal of Luxury Brands." *Fashion Theory* 16, no. 3 (2012): 273–95. https://doi.org/10.2752/175174112X13340749707123

Bedard, Stephanie Anne Nicole, and Carri Reisdorf Tolmie. "Millennials' Green Consumption Behaviour: Exploring the Role of Social Media." *Corporate Social Responsibility and Environmental Management* 25, no. 6 (2018): 1388–96. https://doi.org/10.1002/csr.1654

Better Cotton Initiative. "About Better Cotton." Accessed February 8, 2021. http://bettercotton.org.

Binkley, Sam. "Liquid Consumption: Anti-consumerism and the Fetishized De-fetishization of Commodities." *Cultural Studies* 22, no. 5 (2008): 599–623.

Choi, Yeong-Hyeon, and Kyu-Hye Lee. "Ethical Consumers' Awareness of Vegan Materials: Focused on Fake Fur and Fake Leather." *Sustainability* 13, no. 1 (2021): 1–16. https://doi.org/10.3390/su13010436

Cradle to Cradle Certified. "What Is Cradle to Cradle Certified?" Accessed February 8, 2021. http://www.c2ccertified.org.

Ethical Trading Initiative (EI). "About ETI." Accessed February 8, 2021. https://www.ethicaltrade.org.

Fashion Revolution. "Fashion Transparency Index 2020." Issuu, (2020): 1–70. https://issuu.com/fashionrevolution/docs/fr_fashiontransparencyindex2020.

Fairtrade. "What Is Fairtrade?." Accessed February 8, 2021. http://www.fairtrade.net.

Fair Wear Foundation. "About Us." Accessed February 8, 2021. https://www.fairwear.org.

Giliver, Liam. "Historic: Israel Becomes World's First Country to Ban Fur Sales." Accessed August 17, 2021. https://plantbasednews.org/culture/ethics/israel-becomes-worlds-first-country-to-ban-fur-sales/

Global Organic Textile Standard (GOTS). "The Standard." Accessed February 8, 2021. https://global-standard.org/the-standard.

Harris, Susan M. "Green Tick™: An Example of Sustainability Certification of Goods and Services." *Management of Environmental Quality: An International Journal* 18, no. 2 (2007): 167–78. https://doi.org/10.1108/14777830710725830

Henninger, Claudia E., Panayiota J. Alevizou, and Caroline J. Oates. "What Is Sustainable Fashion? *Journal of Fashion Marketing and Management: An International Journal* 20, no. 4 (2016): 400–16. https://doi.org/10.1108/JFMM-07-2015-0052

House of Commons. "Fixing Fashion: Closing Consumption and Sustainability." Environmental Audit Committee, (2019): 1–73. publications.parliament.uk/pa/cm201719/cmselect/cmenvaud/1952/1952.pdf.

Hughes, Ellie. "How to Tell If a Brand Is Ethical or Not." *The Practivist.* Accessed February 8, 2021. http://thepractivist.com/blog/how-to-tell-if-a-brand-is-ethical-or-not.

International Labour Office. "Global Estimates of Child Labour: Results and Trends, 2012–2016." (2017): 1–68. www.ilo.org/wcmsp5/groups/public/@dgreports/@dcomm/documents/publication/wcms_575499.pdf.

International Labour Organization. "What Is Child Labour?" IPEC. Accessed January 11, 2021. www.ilo.org/ipec/facts/lang--en/index.htm.

Kuhlman, Tom, and John Farrington. "What Is Sustainability?" *Sustainability* 2, no. 11 (2010): 3436–48. https://doi.org/10.3390/su2113436

London Design Collective. "Why Vegan Fashion Has Grown in Popularity." Accessed August 17, 2021. https://londondesigncollective.com/fashion/why-vegan-fashion-has-grown-in-popularity/.

Lundblad, Louise., and Iain A. Davies. "The Values and Motivations Behind Sustainable Fashion Consumption." *Journal of Consumer Behavior* 15, no. 2 (2016): 149–62. https://doi.org/10.1002/cb.1559

Lundahl, Outi. "Dynamics of Positive Deviance in Destigmatisation: Celebrities and the Media in the Rise of Veganism." *Consumption Markets & Culture* 23, no. 3 (2020): 241–71. https://doi.org/10.1080/10253866.2018.1512492

Maiti, Rashimila. "Fast Fashion: Its Detrimental Effect on the Environment." *Earth.Org,* January 29, 2020. earth.org/fast-fashions-detrimental-effect-on-the-environment/.

Mao, Jin. "Social Media for Learning: A Mixed Methods Study on High School Students' Technology Affordances and Perspectives." *Computers in Human Behavior* 33, (2014): 213–23. https://doi.org/10.1016/j.chb.2014.01.002.

Mehta, Angeli. "Beyond Recycling: Putting the Brakes on Fast Fashion." *Reuters Events*, April 28, 2019. www.reutersevents.com/sustainability/beyond-recycling-putting-brakes-fast-fashion.

Mhugos. "Zara Clothing Company Supply Chain." *SCM Globe*, January 4, 2020. www.scmglobe.com/zara-clothing-company-supply-chain/.

Montanari, Giulia. "Mary Ellen Wilson: When Abused Children Had Fewer Rights Than Pets." *Medium*, August 12, 2020. medium.com/history-of-yesterday/mary-ellen-wilson-the-abused-child-rescued-by-animal-rights-activists-9dfb5b1f100a.

Moulds, Josephine. "Child Labour in the Fashion Supply Chain." *The Guardian and Media*, accessed January 11, 2021. labs.theguardian.com/unicef-child-labour/.

Organic Content Standard (OCS). "What Is the OCS?." Accessed February 8, 2021. http://textileexchange.org.

PETA. "PETA Pushes Fashion Brands to Move Beyond Using Fur—and Wins." Accessed August 17, 2021. https://www.peta.org/features/fur-free-companies-brands-that-banned-fur/.

Population Council. "Vienna Declaration on Human Rights." *Population and Development Review* 19, no. 4 (1993): 877–82. https://doi.org/10.2307/2938429

Sexton, Alexandra E., Tara Garnett, and Jamie Lorimer. "Vegan Food Geographies and the Rise of Big Veganism." *Progress in Human Geography* 48, no. 2 (2022): 605–28. https://doi.org/10.1177/03091325211051021

Shephard, Aelsea, and Sanjukta Pookulangara. *The Slow Fashion Process: Rethinking Strategy for Fast Fashion Retailers*. Florida: CRC Press, 2013.

SOMO. "Fact Sheet Child Labour in the Textile & Garment Industry." (2014): 1–10. www.somo.nl/wp-content/uploads/2014/03/Fact-Sheet-child-labour-Focus-on-the-role-of-buying-companies.pdf.

Stanton, Audrey. "What Is Fast Fashion, Anyway?" The Good Trade. Accessed February 8, 2021. www.thegoodtrade.com/features/what-is-fast-fashion.

Stringer, Tara., Gary Mortimer, and Alice Ruth Payne. "Do Ethical Concerns and Personal Values Influence the Purchase Intention of Fast-Fashion Clothing?" *Journal of Fashion Marketing and Management: An International Journal* 24, no. 1 (2020): 99–120.

Su, Jin, and Aihwa Chang. "Factors Affecting College Students' Brand Loyalty toward Fast Fashion." *International Journal of Retail & Distribution Management* 46, no. 1 (2018): 90–107.

Tariq, Mehreen Ghani. "Inside the Ugliness of the Fast Fashion Industry." *The Medium*, July 9, 2020. https://medium.com/maverickyouth/inside-the-ugliness-of-the-fast-fashion-industry-ac40f6a24e01.

Thiele, Leslie P. *Sustainability*. Cambridge: John Wiley & Sons, 2013.

Tobler, Christina, Vivianne Visschers H. M., and Michael Siegrist. "Addressing Climate Change: Determinants of Consumers' Willingness to Act and Support Policy Measures." *Journal of Environmental Psychology* 32, no. 3 (2012): 197–207. https://doi.org/10.1016/j.jenvp.2012.02.001

Tokatli, Nebahat. "Global Sourcing: Insights from the Global Clothing Industry—the Case of Zara, A Fast Fashion Retailer." *Journal of Economic Geography* 8, no. 1 (2008): 21–38. https://doi.org/10.1093/jeg/lbm035.

United Nations. "17 Goals to Transform Our World." United Nations Global Impact. Accessed February 8, 2021. https://www.unglobalcompact.org/sdgs/17-global-goals.

———. "Overview." United Nations: Peace, Dignity and Equality on a Healthy Planet. Accessed February 8, 2021. www.un.org/en/sections/about-un/overview/index.html.

———. "History of the United Nations." United Nations: Peace, Dignity and Equality on a Healthy Planet." Accessed February 8, 2021. www.un.org/en/sections/history/history-united-nations/index.html.

———. "Chapter 1." United Nations: Peace, Dignity and Equality on a Healthy Planet. Accessed February 8, 2021. www.un.org/en/sections/un-charter/chapter-i/index.html.

United Nations Human Rights. "Vienna Declaration and Programme of Action." OHCHR. Accessed February 8, 2021. www.ohchr.org/en/professionalinterest/pages/vienna.aspx.

———. "What are human rights." OHCHR. Accessed February 8, 2021. https://www.ohchr.org/en/issues/pages/whatarehumanrights.aspx.

Woodside, Arch G., and Monica B. Fine. "Sustainable Fashion Themes in Luxury Brand Storytelling: The Sustainability Fashion Research Grid." *Journal of Global Fashion Marketing* 10, no. 2 (2019): 111–28. https://doi.org/10.1080/20932685 .2019.1573699

Chapter Nine

Making the Material Turn

A Pedagogical Approach on Postcolonial, Social, and Ecological Issues in Amitav Ghosh and Arundhati Roy's Essays and Fiction

Suhasini Vincent

Postcolonial ecocriticism examines the relationship between humans and nonhuman communities and strives to imagine new ways in which these ecologically connected entities can coexist, transform and reconfigure themselves.[1] Even though postcolonial environmental issues are often linked to the colonizer's invasion, settlement and depletion of natural resources, forced migration of peoples, and the fretting of animals and plants throughout the European empires, the ecosystem challenges that are faced by postcolonial countries today are due to nonegalitarian government policies and corporate-capitalist dominance. Students today are aware of the North-South divide and wonder if there are means of reconciling the Northern environmentalism of the rich and the Southern environmentalism of the poor. History textbooks reveal the ever-present ecological gap between the colonizer and colonized and the disparities that continue to exist.

Ecocritical readings as seen through the lens of Amitav Ghosh[2] and Arundhati Roy[3] reveal the need to bring postcolonial and ecological issues together and challenge imperialist modes of social environmental supremacy.

Both writers have considered the complex interplay of politics and corporate capitalism in the use of water, land, energy, and habitat. Their fictional and nonfictional ecocritical writing trace the social, historical, political, economic, and material coordinates of forests, rivers, bioregions, and species. It is

177

interesting for a student to explore the myriad relationships between material practices and cross-cultural contexts in the postcolonial world. For example, an American student's curiosity would be awakened by how day-to-day life is lived in postcolonial India. This chapter will reveal how futures are governed in the developing world. An ecocritical reading of works by Ghosh and Roy will examine how the policies of decision-makers and ideas from proponents of climate change can lead to a transference from a "red" to a "green" politics. The field of ecocriticism is large as it encompasses domains of economics, anthropology, history, geography, geopolitics, and environmental science. Thus, this chapter will instigate students in high school and college levels, as well as researchers in different disciplines, to analyze texts by considering how the authors advocate different means of dwelling authentically and responsibly on our planet Earth.

Ghosh's *The Great Derangement: Climate Change and the Unthinkable* aims at exploring how long-term ecological crises have been ignored in serious fictional writing.[4] Ghosh wisely reflects on how "different modes of cultural activity: poetry, art, architecture, theater, prose fiction . . . have responded to war, ecological calamity, and crises of many sorts: why then, should climate change prove so peculiarly resistant to their practices?"[5] He muses further, "What is it about climate change that the mention of it should lead to the banishment from the preserves of serious fiction? And what does this tell us about culture writ large and its patterns of evasion?"[6] He foretells a scenario of a world where cities' forests like the Sundarbans would be replaced by seascapes; megalopolises would become uninhabitable; future generations would try to find answers in literature and art for the damaged world they had unwittingly inherited. He imagines future generations concluding that our present generation is rather deranged and oblivious to the hazards, risks, vulnerabilities, and perils of climate change:

> In a substantially altered world, . . . when readers and museum goers turn to art and literature of our time, will they not look, first and most urgently, for traces and portents of the altered world of their inheritance? And when they fail to find them, what should they—what can they—do other than to conclude that ours was a time when most forms of art and literature were drawn into the modes of concealment that prevented people from recognizing the realities of their plight . . . this era which so congratulates itself on its self-awareness, will come to be known as the time of the Great Derangement.[7]

In this work of nonfiction, Ghosh probes into possible reasons for the postcolonial writer's imaginative failure in the face of overwhelming evidence. He examines the inability of the present generation to grasp the violent scale of climate change and posits that this is reflected in the literature of our time, in the recording of history, and in the political ambience of our day. Ghosh posits that the extreme nature of today's climate changes has resulted in making writers immune and resistant to contemporary modes of thinking and imagination. He points out that certain phenomena like destructive storms and meandering rivers do not figure in serious literary fiction but get relegated to other genres like science fiction or fantasy. Ghosh suggests that politics has suffered the same fate as literature and has become a matter of personal moral reckoning rather than an arena of collective action. He further argues that to limit fiction and politics to individual moral adventure comes at a great cost. Through his writing Ghosh has shown how fiction is the best of all cultural forms to voice ecocritical awareness.

Arundhati Roy's political nonfiction[8] has often been viewed with a disdainful eye[9] and considered as "objectionable writings."[10] In her essay,[11] Roy voices dissent against repression, globalization, economic progress, environmental exploitation, and dams proposed in the name of greater common good. Roy's political essays insist on the need to make a "material turn" and considers how materialistic activity in the name of progress affects human and nonhuman environments. The term "material turn" is used by Serenella Iovino in her work *Material Ecocriticism* in which she describes an enterprise of writing that is replete with "a material mesh of meanings, properties and processes, in which human and non-human players are interlocked in networks that produce undeniable signifying forces."[12] Roy's essays challenge government policies, discuss varying viewpoints of global and local concerns in India, criticize corporate philanthropy, and propose a new ecocritical perspective. Through her essays Roy expresses distrust of materialistic forces and encourages readers to take part in the new emerging paradigm of making a "material turn," thus considering possible ways of avoiding the depletion of our planet's resources through ecocritical advocacy and aesthetics. But one wonders why she writes about climate change in nonfictions, opting for the mode of the essay. Amitav Ghosh echoes the same idea by insisting that this practice is not due to a lack of information—"A case in point is the work of Arundhati Roy: not only is she one of the finest prose stylists of our time, she is passionate and deeply informed about climate change. Yet all her writings on these subjects are in various forms of nonfiction."[13] Ghosh also provides potential reasons for this sleight of writing by contending that "the discrepancy is not the result of personal predilections: it arises out of peculiar forms of resistance that climate change presents to what is now regarded as serious fiction."[14] In her "Arthur Miller Freedom to Write" lecture, Roy poses a series of questions on what it means to be a writer in today's world:

So, as we lurch into the future, in this blitzkrieg of idiocy, Facebook "likes," fascist marches, fake-news coups, and what looks like a race toward extinction—what is literature's place? What counts as literature? Who decides? Obviously, there is no single, edifying answer to these questions.[15]

By musing on who decides, she concludes that there are no enlightening answers to these questions. She posits that the time is right to think together about a place for literature and the role it will play in climate change. In the same lecture, Arundhati Roy muses on what it means to be an "activist-writer" in a world where the delicate web of interdependence of Man and Nature is dictated by capitalism and international treaties. Roy's essays in India are often regarded with a baleful eye as she disagrees with political decision-making, arouses dissent among the youth and writes forcefully on topics other writers do not explore. She acknowledges that some did not count her nonfiction as writing:

> The writing sat at an angle to what was conventionally thought of as literature. Balefulness was an understandable reaction, particularly among the taxonomy-inclined—because they couldn't decide exactly what this was—pamphlet or polemic, academic or journalistic writing, travelogue, or just plain literary adventurism? To some, it simply did not count as writing. "Oh, why have you stopped writing? We're waiting for your next book."[16]

Though some considered her a mere "pen for hire,"[17] others were open to the call of change as her essays in the form of pamphlets were translated into other Indian languages, distributed freely in villages that were under attack, on university campuses where students were realizing that they were being lied to. The readership concerned those on the frontlines who understood her brand of literature.

Both Arundhati Roy and Amitav Ghosh believe that literature contributes to change and that it is built by writers and readers. Literature in their opinion occupies a fragile place that needs constant shelter and nurturing. With a pedagogical approach that harnesses the attention of readers, arouses ecocritical awareness of climate change and prepares the youth of today to conserve our planet for the next generations, the present generation may not be doomed to the derangement Ghosh laments.

INTRODUCTION: EXPLORING THE
WRITER'S LIVED EXPERIENCE

Postcolonial writers mine their own lived experiences to create a fictional ground that reflects the reality of their times. Amitav Ghosh has a doctorate in social anthropology[18] and his novels bear ample evidence of the ethnographic research and fieldwork he has conducted. His fictionalized narratives often draw from his own personal experience of migration and displacement. Arundhati Roy, an architect[19] involved in political activism, is a spokeswoman of the downtrodden and vehement critic of neoimperialism,[20] fascism, corruption, and communalism. Her essays aim at laying bear the shortcomings in governmental decision making and inspiring the youth of today react. While using the works of Ghosh and Roy in a classroom, teachers should guide students to explore the fictional space by encouraging an analysis of select quotations referred to in this chapter. Teachers could choose a section for detailed study and surf the authors' websites[21] which contain excerpts of newspaper reviews and authors' notes. They could complete a detailed reading of a work and analyze it through the following three-step lens: figuration of the natural and material lay of the land; configuration of the writer's inherited experience in the form of stories, myths, legends, culture and art; and the reconfiguration of the fictional or nonfictional space to inspire young readers to engage in the politics of climate change.

Mining and Recasting Childhood Stories for the Fictional Ground

In *The Great Derangement*, Amitav Ghosh mines his own experience, draws inspiration from stories of his childhood and reconfigures the fictional ground. As he notes,

> No less than any other writer have I dug into my own past while writing fiction. . . . In essence, narrative proceeds by linking together moments and scenes that are in some way distinctive or different: these are, of course, nothing other than instances of exception . . . It is through this mechanism that worlds are conjured up, through everyday details, which function "as the opposite of narrative." It is thus that the novel takes its modern form, through "the relocation of the unheard-of toward the background . . . while the everyday moves into the foreground."[22]

He recalls childhood family accounts of his ancestors who had been ecological refugees displaced from their roots on the banks of the River Padma[23] in the mid-1850s. When the mighty river on a whim decided to change course, the few vagrant inhabitants traversed forests and dry land moving westward to settle once again on the banks of another civilization-housing river, the Ganges.[24] Ghosh warns us that the river, a stable presence in the lives of his forefathers, had transformed into a meandering force that could not be relied on, nor taken for granted like the air we breathe. In his novel *The Hungry Tide*, Ghosh writes of the fickle nature of the landscape where mangrove forests of the Sundarbans[25] appear and disappear, merge and submerge, surprise and disrupt human lifestyles. He muses in *The Great Derangement*, "Even a child will begin a story about his grandmother with the words: 'in those days the river wasn't here, and the village was not where it is. . . . '"[26] The fictional ground of *The Hungry Tide* is set in the fickle tidal landscape, or *bhatir desh*,[27] of the Sundarbans where the passage of the ebb tide leaves an ever mutating and unpredictable terrain with "no borders to divide fresh water from salt, river from sea as the water tears away entire promontories and peninsulas; at other times it throws up new shelves and sandbars where there were none before."[28] In this indeterminate fluid fictional space of the Sundarbans, Ghosh takes the reader-voyager on a literary journey across the Gangetic delta. The voyage reveals the varied quests, travels, expeditions, and voyages of the protagonists in the econarrative. The mighty River Hooghly[29] dictates the literary journey in the novel as it meanders, changes course, reshapes land before sunrise, and reconfigures them with new paths before sunset. During its ever-shifting course, the River Hooghly intertwines with the River Meghna,[30] thus setting the theme of transformation and displacement for the human and animal inhabitants who adapt to the capricious river trajectories and deal with the challenging task of naming new, fresh, ever-emerging islands each day.

The character of Fokir, a local fisherman, recites a "legend passed on from mouth to mouth, and remembered only in memory."[31] The legend of Bon Bibi[32] features the relationship between human settlers and tiger predators in the Sundarbans. It describes an epoch when Bon Bibi had the divine task of rendering the Sundarbans fit for human inhabitation. In a battle between "good" and "evil," Bon Bibi emerges triumphant, divides the country of eighteen tides into two inhabitable zones for "humans" and for "demon-tiger" hordes of Dokhin Rai,[33] thus changing the time-set laws of the "survival of the fittest" to a new-named "law of the forest." The rich and greedy are aptly punished while the poor and righteous reap just rewards. In Fokir's river song, Dokhin Rai takes the form of the Ganges tiger with an insatiable craving for human flesh, and the fearsome tiger stalks Dhukey, a fisher boy, thus chanting an ever-living myth to affirm the hostilities between "death-bearing," "demon" tigers versus "precious," "precarious" humans. Through the inclusion of the myth of Bon Bibi, Ghosh sets his novel in a twofold time zone of the present and the past that is influenced by a mythical time of incessant mutiny between "humans" and "animals," both species being constantly threatened by the vagaries of the river and the sea tide. By using the recurring chant of the Bon Bibi song, Ghosh highlights how myth influences and affects people's existence in the Sundarbans. Through the inclusion of the myth of Bon Bibi, he portrays the Sundarbans as a living entity endowed with the capacity to nurture "human" and "animal" protagonists who claim equal land, space and territory thanks to an entitlement to land that can be traced back to roots in myth. Creation myths, nature fables, and family stories exist in many cultures and families. Teachers can ask students to engage in an investigation or research project of such myths, legends, anecdotes, and stories that may have been passed down to them through their parents or grandparents. Like Ghosh who remembers his grandmother's accounts of meandering rivers and migration induced by a quest for other civilization-housing river zones, students may unearth similar tales of lost forests, farmlands that now accommodate skyscrapers, changing weather conditions, disappearing plant and animal life, and changing relationships with our planet Earth.

Figuring the Material Lay of the Land and Mining the Arts for Material Gain

In her second novel *The Ministry of Utmost Happiness*,[34] Roy writes "People—communities, castes, races and even countries—carry their tragic histories and their misfortunes around like trophies, or like stock, to be bought and sold on the open market."[35] In her essay "Power Politics," Roy shows how the ruling elite in India, bureaucrats and multinational companies, welcome foreign dignitaries and present India as an investment conducive haven. Roy describes how Delhi was transformed, scrubbed, and cleaned to welcome Bill Clinton and his delegates in March 2000: "whole cities were superficially spruced up. The poor were herded away, hidden from the presidential gaze. Streets were soaped and scrubbed and festooned with balloons and welcome banners."[36] In *Capitalism: A Ghost Story,* Roy criticizes the Indian government strategy of creating an aura of "Good Investment Climate"[37] and describes how Rand and Enron ended up owning Indian earth, air, and water. She questions the motives behind the drive to beautify Delhi for the Commonwealth Games when ephemeral laws were passed to present Delhi as a beautiful capital without any marring disfigurement. She describes how

street vendors disappeared, rickshaw pullers lost their licenses, small shops and businesses were shut down. Beggars were rounded up, tried by mobile magistrates in mobile courts, and dropped outside the city limits. The slums that remained were screened off, with vinyl billboards that said DELHIciously Yours.[38]

Through metaphors of cleansing, Roy highlights how the capital city was purged of unwanted elements that marred the perfect scene staged to impress foreign investors. By eliminating the presence of street vendors, beggars, slums, and the poor, Delhi was reconfigured into a city that was inviting for foreign investment.

Further, Roy remarks on how main mining conglomerates mine the arts—film, art installations and literary festivals—to gain popularity as patrons of artistic and literary enterprise. She gives instances of how they exploit the arts for material gain: the bauxite mining conglomerate Vedanta sponsors a film competition named "Creating Happiness" encouraging the youth to make films on sustainable development[39]; Jindal Group,[40] the stainless-steel giant, has a contemporary art magazine and supports artists who work with the same medium of stainless steel; Indian energy giant Essar, despite its mining scandal,[41] boasts of being the principal sponsor of the Tehelka Newsweek Think Fest[42] with its mind-stirring discussions by major writers and activists; the involvement of Tata Steel[43] and Rio Tinto[44] in the Jaipur Literary Festival[45] where the right to free speech in the style of Salman Rushdie is much publicized as voices "talking about the house, about everyone in it, about everything that is happening and has happened and should happen."[46][47] Despite the public display of embracing the arts, these miners of resources both natural and creative ignore "journalists, academics and filmmakers working on subjects unpopular with the Indian Government."[48]

Exploiting India for Material Gain

Postcolonial writing aims at revealing areas of colonization and its impact on the contemporary postcolonial world. The fiction and nonfiction of Amitav Ghosh and Arundhati Roy reveal the ambivalent themes and issues of colonial legacy that have shaped the so-called developing BRIC[49] countries of today. Ghosh's fiction examines the material gain of the colonists and analyzes the dynamics of colonial trade, including its environmental impact on the indigenous population of colonial times. Arundhati Roy's fiction bears, on the contrary, less reference to colonial times, but gives a stark account of contemporary power politics, where richer countries dominate in international forums as experts on environmental development. She highlights also the imbalance of power politics in India where privatization has resulted in material wealth being held in the hands of a few mega corporations that thrive on material gain. While presenting the work of postcolonial authors, it is thus imperative to have a knowledge of the colonial past and engage in an exploration of the Opium Triangular Trade conducted by the British East Indian Company, which can be an eye-opener for students. The contemporary power politics in the Indian postcolonial context may be investigated by researching each one of the mega corporations mentioned in the ensuing sections.

The Opium Triangular Trade of the British East India Company

In the Ibis trilogy, Ghosh gives an account of the materialistic intent behind the British East India Company's Opium Trade in India and their trade settlement in the Chinese province of Canton in the nineteenth century. Ghosh's econarrative depicts the resistance mounted by the Chinese authorities in protest against the lucrative British trade in the Golden Triangle.[50] The Golden Triangle refers "to a broadly triangular area with vertices with Burma, Laos and Thailand, where opium production was concentrated."[51] Ghosh's fictions highlight the empire-sized fortunes earned by the materialistically oriented British traders and pictures the suffering faced by the indigenous Indian and Chinese populations who stagger under the stupor of the potent somniferous drug of opium.

In the first novel of the Ibis trilogy, *The Sea of Poppies*, the Black Water of the Sundarbans in the novel is seen as an expanse of water without "a boundary, a rim, a shore, to give it shape, and hold it in place."[52] Not only is the countryside blanketed with the parched remains of the poppy harvest, the debris of poppy leaves, dumped on the shore makes its way to the Black Andaman Sea.[53] Through the drastic change in the landscape from rich and fertile irrigated plains to that of swamps and marshes with no potable water, Ghosh highlights the effects of greedy colonial enterprise on nature. In the second novel, *River of Smoke*, Ghosh's narrative takes the action forward from the cities, harbors, and plains of India to the Chinese trading outpost and opium destination of Canton. In the tiny foreign enclave of Fanqui-town,[54] the Cantonese outpost is populated by traders of the British East India Company[55] and surrounded by a flotilla of boats that ferry smuggled goods and serve as eating and pleasure houses. The West's fascination for the East as seen in the Orientalist-initial quest for spices like nutmeg, cloves, pepper, coffee, cacao, sugar, and tea lead eventually to the discovery of the addictive narcotic drug "opium." In *Opium, Empire, and the Global Political Economy*, Carl Trocki posits that "the British Empire, the opium trade, and the rise of global capitalism all occurred together. . . . Gold and silver from the West crossed the Atlantic or the Pacific, it ultimately found its way to Asia (east of Suez) to purchase the 'riches of the East' and to allow the otherwise deprived inhabitants of the northwest Eurasian peninsula to share in the fabled Oriental splendours."[56] This "ineradicable distinction between Western superiority and Oriental inferiority"[57] described by Edward Said[58] is evident in Ghosh's postcolonial textual response in the Ibis Trilogy where the margins write back and lay bare materialistic colonial intent of the past by reconfiguring and reliving the opium trade in the fictional space.[59] In *Flood of Fire*, the final book in the

Ibis trilogy, Ghosh describes how the British mobilized force on a large scale, unleashing the firepower of their advanced warship called *Nemesis*:[60]

> Baboo Nob Kissin raised a hand to point to the Nemesis, which was steaming past the burning forts, wreathed in dark fumes. Dekho—look: inside that vessel burns the fire that will awaken the demons of greed that are hidden in all human beings. That is why the British have come to China and Hindustan: these two lands are so populous that if the greed is aroused, they can consume the whole world. Today that great devouring has begun. It will end only when all of humanity, joined together in a great frenzy of greed, has eaten up the earth, the air, the sky.[61]

The novel is aptly named a *Flood of Fire*, as fire engulfs the Indian vessels. The incessant flood of fire prompts the Indian soldier Kesri to reflect, "So much death; so much destruction—and that too visited upon a people who had neither attacked or harmed the men who were so intent on engulfing them in this flood of fire."[62]

Ghosh's fictional enterprise of reliving the colonial trade is typical of what Ashcroft, Griffiths, and Tiffin claim to be "the rereading and the rewriting of the European historical and fictional record" (that) "is a vital and inescapable task at the heart of the postcolonial enterprise."[63,64] He depicts how the British citizens of a global imperium dominate the world through a false doctrine of free trade. Ghosh shows that "despite all their cacklings about Free Trade, the truth was that their commercial advantages had nothing to do with markets or trade or more advanced business practices—it lay in the brute firepower of the British Empire's guns and boats."[65] During this era of triangular trade that enabled the three-legged journey of exchanging slaves in Africa for guns and brandy, the Middle Passage across the Atlantic to sell the slaves in the West Indies and North America, and the final cargo transportation of rum and sugar to England, opium served as the means of conducting business. It should be remembered that colonial trade conducted by these British opium barons contributed to colonial empire-building, large-scale opium production, and trading in the colonies, which led to the creation of the Golden Triangle[66] of Burma, Laos, and Thailand and the Golden Crescent of Afghanistan, Iran, and Pakistan.[67]

Privatization and the Dynamics of "Power Politics"

In *Capitalism: A Ghost Story*, Roy reminds readers that 800 million impoverished Indians "live side by side with spirits of the nether world, the poltergeists of dead rivers, dry wells, bald mountains and denuded forests."[68] This enumeration of dead resources essential for the survival of mankind serves to highlight significant figures: while 800 million impoverished Indians live on an average of twenty rupees per day, the top 100 richest industrialists in India possess wealth that amounts to one-fourth of the country's GDP of $300 billion. Material wealth in India is concentrated in the hands of a few. Roy juxtaposes her enumeration of depleted resources with the assets of India's richest billionaire Mukesh Ambani, whose twenty-seven-story residence houses "three helipads, nine lifts, hanging gardens, ballrooms, weather rooms, gymnasiums, six floors of parking, and the six hundred servants."[69] Ambani's Reliance Industries Limited is typical of the market scenario in India where corporations like Tatas, Jindals, Vedanta, Mittals, Infosys, and Essar are engaged in a race for supremacy in India's private sector. As Roy notes,

> they own mines, gas fields, steel plants, telephone, cable TV, and broadband networks, and run whole townships. They manufacture cars and trucks, own the Taj Hotel chain, Jaguar, Land Rover, Daewoo, Tetley Tea, a publishing company, a chain of bookstores, a major brand of iodized salt, and the cosmetics giant Lakme. Their advertising tagline could easily be: You Can't Live without Us.[70]

Roy labels this corporate race as "Gush-Up Gospel" and posits that "the more you have, the more you can have."[71] Roy details that the motive behind the foreign visits was to sweet-talk India into importing products which the country could manufacture on its own. In her essays, Roy warns readers that foreign investors were stalking big, varied game in the form of dams, mines, telecommunication, public water supply, and the dissemination of knowledge.

Roy notes that in the present trend of privatization, "India's new mega-corporations Tatas, Jindals, Essar, Reliance, Sterlite, are those who have managed to muscle their way to the head of the spigot that is spewing money extracted from deep inside the earth."[72] She contends that massive corporations, both multinational and domestic, are the principal agents of the concentration of wealth, and exploit the government's present policy of creating Special Economic Zones to have access to precious land and resources, resulting in the dispossession of millions. Whole-scale privatization has spurred massive corruption and resulted in massive displacement of poor people who were promised inexistent jobs. She explains that with reference to water politics where "all over the world, weak, corrupt, local governments have helped Wall Street brokers, agro-business corporations and Chinese billionaires to amass huge tracts of land."[73] This is a reference to Roy's essay "Power Politics," where she comments on the farce behind the meeting of International World Water Forum in March 2000 at The Hague. While 3,500 proponents like bankers, businessmen, politicians, economists, and planners pressed for the privatization of water, a handful of activist-opponents expressed dissent against the appropriation of a national resource by private multinationals. While the forum engaged in upholding false interest in issues regarding "women's empowerment, people's participation, and deepening democracy,"[74] the need to preserve the life-giving resource was pushed to the background. She exposes the existence of a lobbying effort at The Hague, where cliques of consultants specialized in preparing dossiers for the third world presented concocted data and virtual facts and figures. She states that they "breed and prosper in a space that lies between what they say and what they sell."[75] She thus shows how privatization thrives through the incomprehensible jargon of government dossiers and hostile takeover bids. Roy explains that privatization of India's mountains, rivers, and forests necessarily entails war, displacement, and ecological devastation. She refers to the signing of MOUs by the state governments of Chhattisgarh, Orissa, and Jharkhand[76] in 2005 when private corporations mined material gain worth trillions of dollars from mining bauxite, iron ore, and other minerals for "a pittance, defying even the warped logic of the Free Market. (Royalties to the government ranged between 0.5% and 7%)."[77] Her essay is written with the intention of protecting India's power reserves and seeks to reveal the dynamics of power politics to the world at large.

ENCOURAGING SOCIAL AND ENVIRONMENTAL ADVOCACY IN LITERATURE

Postcolonial ecocritical writing focuses on the political and geopolitical tensions that exist between the colonizer and the colonized during the twentieth century. Amitav Ghosh's fictional account reveals the efforts of colonists to create material prosperity detrimental to the colonized; on the other side, he configures how ecofriendly communities sprout in the fictional space to discover, identify, protect, and conserve threatened interconnected ecosystems and ecosocieties. While exploring Ghosh's environmental advocacy, students should be encouraged to analyze how his fictional enterprise falls within the sphere of postcolonial ecocriticism that considers the problems of conserving biodiversity and distrusts the grandeur of empty materialistic quests. An ardent activist, Arundhati Roy's writing is a result of her field work and involvement in environmental issues. She criticizes the Indian government's strategy of engaging in development at the cost of uneducated villagers' losing their livelihood and land. While discovering Roy's account of postcolonial India, teachers should encourage students to analyze how literature, both fictional and nonfictional, hold the key to environmental advocacy as it reveals the author's perspective of how development is not always for the greater common good. Teachers may encourage students to investigate their own country's governmental practices, policies, and decision-making. There may be recorded instances of how a government's environmental strategies created more harm than good. The goal is not to resurrect the forgotten stories of history to create shame or throw blame on students for inaction but to encourage a more contextualized understanding of injustice. This pedagogical approach aims at discovering the alignment of a green paradigm through an econarrative and environmental critical advocacy. This attempt to unite fictional aesthetics and advocacy is typical of postcolonial ecocriticism that seeks to advocate social and environmental justice in today's postcolonial world.

Establishing a Green Paradigm through an Econarrative

In Ghosh's econarrative *The River of Smoke*, the troubled waters of the South China Sea witness the tryst between two storm-tossed vessels: the *Anahita*, a sumptuously built cargo ship laden with opium, and the *Redruth*, a two-mast vessel with a Cornish botanist and his assistant, Paulette, an orphan who collects rare flora during the stormy journey. Paulette catalogues the plants of Bengal and contributes to the body of collected knowledge called the *Materia Medica*.[78] She joins forces with the protagonist, a famous plant hunter, Fitcher Penrose, to search for a rare camellia[79] and seek newer vistas through botanical exploration. Ghosh also pens the route of pilgrimage undertaken by early horticulturists to the Pamplemousses garden in his econarrative. He writes of the existence of a chaotic botanical garden where a wild and tangled muddle of greenery showed the existence of a primeval jungle "where African creepers were at war with Chinese trees, nor one where Indian shrubs and Brazilian vines were locked in a mortal embrace. This was a work of Man, a botanical Babel."[80] Thus, Ghosh reconfigures new spaces of postcolonial identity through an econarrative that enables the reader to identify himself with the central premise of articulating resistance against materialistic forces. Paulette and Penrose voice the need to protect greenery and nature from forces that strive to destroy the landscape. As a child of nature, Paulette had been taught by her father, Pierre Lambert, to love nature and consider it as a kind of spiritual striving whereby the quest was to comprehend the inner vitality of each species:

If botany was the Scripture of this religion, then horticulture was its form of worship: tending a garden was, for Pierre Lambert, no mere matter of planting seeds and pruning branches—it was a spiritual discipline, a means of communicating with forms of life that were necessarily mute and could be understood only through a careful study of their own modes of expression—the languages of efflorescence, growth and decay: only thus he had taught Paulette, could human beings apprehend the vital energies that constitute the Spirit of the Earth.[81]

During the passage through the North China Sea, Paulette identifies a large variety of plants and tends them like a priestess performing a spiritual ritual. Ghosh lays stress on the interconnected nature of different life forms and Paulette's quest to name unnamed flora from Chinese territory that can be considered as a creative endeavor to consider how these ecologically connected groups can be creatively transformed. The mundane tasks of planting, pruning, and watering them become acts of discipline and a means of communication with mute forms of existence that manifest different kinds of vital energies that constitute the spirit of the Earth. She carefully observes the procedures and protocols on board during times of storms. At the same time the ecoconservationists in the fictional space have to deal with the discontent of the seamen on board who regard the plants as threats to their existence and deny them water in times of scarcity or empty the pots of precious water when menaced by storms. The reader encounters a multiplicity of voices that express the problems that the world faces today and discovers a plethora of issues that speak of the need to assert a green paradigm[82] free of the stamp of lucrative colonial trade.

Fighting against Dams in the Name of Greater Common Good

In *Capitalism: A Ghost Story*, Roy criticizes the building of dams in the name of common good. She wonders, "How can they stop a dam?"[83] referring to the Kalpasagar project[84] that has ambitious plans of supplying water to SIR and SEZ projects that Roy labels as a self-governed corporate dystopia of "industrial parks, town-ships and mega-cities."[85,86] In an earlier political essay-manifesto titled "Greater Common Good," Roy defends the Adivasi tribal people,[87] who lost their homes and livelihood to the construction of the Sardar Sarovar Dam in the Narmada valley.[88] This essay's preface is dedicated to the River "Narmada and all the life she sustains."[89,90] The opening lines—"If you should suffer you should suffer in the interests of the country"[91]—are a nod to a speech made by India's first Prime Minister Jawaharlal Nehru[92] who described dams as temples of modern India while addressing homeless villagers who had been displaced by the Hirakud dam construction in 1948.[93] She reminds readers that displacement and evacuation of villagers will be inevitable once again with the Kalpasagar dam. Her essays aim to "puncture the myth about the inefficient, bumbling, corrupt, but ultimately genial, essentially democratic, Indian state."[94] In the power struggle between the protagonists of common good who support development ventures and adversaries who favor a pre-industrial dream, she accuses both groups of resorting to "deceit, lies, false promises and increasingly successful propaganda"[95] to configure a sense of false legitimacy. She implies that the "Iron Triangle" composed of politicians, bureaucrats, and dams work hand in glove with British consultants of the world to devise environmental impact assessments (EIAs) that mask and hide the unavailability of water statistics, the destruction of flora and fauna, and the mass exodus of uneducated villagers. Roy accuses the Indian government of violating the human rights to a normal standard of life of innocent people as they "stand to lose their homes, their livelihoods, their gods and their histories."[96] She states that this quest for modernity, as demonstrated in the creation of the Dholera SIR,[97] where there is the risk of an extinction of rare fish species, is typical of the selfish human decision to survive at the expense of wildlife. She insists on the need to find means of "exploiting nature while minimizing non-human claims to a shared earth."[98,99] Roy's essay, with its hidden agenda of social and environmental advocacy, is imaginative and serves as "a catalyst for social action and [. . .] a full-fledged form of engaged cultural critique."[100] With time running out, Roy insists that words can prove to be the best arms to protect India's chemical polluted rivers. Her fight against the dam does not involve just the tryst with the Iron Triangle, but also encompasses the struggle to preserve a whole ecosystem of cropping and breeding patterns of humans and animal species. Through Roy's account the reader discovers a clear lay of the land defining "what happened where and when and to whom."[101] She alerts readers that the Narmada Valley containing fossils, microliths of the Stone Age, and the history of the Adivasis was doomed in the name of common good.

CONCLUSION: MAKING THE MATERIAL TURN

This pedagogical approach considers the theoretical framework of the material turn which conceives matter as an agentic force with an effective and transformative power over human and nonhuman environments. Amitav Ghosh and Arundhati Roy explore the problems of conserving environmental biodiversity, distrust materialistic intent that crushes ordinary people, and take part in advocating the new emerging dynamics of making a material turn. Thus, their texts offer possible ways of analyzing language and reality, human and nonhuman life, mind and matter, without falling into dichotomous patterns of thinking. Both authors probe into the reasons for the postcolonial writer's imaginative failure in the face of global warming. They insist on the need for the younger generation to take part in environmental politics and posit that this should be reflected in the literature of our age, in historical accounts, and in political decision-making of contemporary postcolonial governments. Ghosh and Roy resist stasis[102] and campaign for change by advocating new contemporary modes of thinking and imagination. This pedagogical approach of analyzing the material turn will enable students and teachers to contemplate how Ghosh's and Roy's ecocritical writing is the best of all cultural forms to voice ecocritical awareness.

Econarrative vs. Ecomaterialism

Ghosh's narrative in the Ibis trilogy strives to remember the materialist colonial past, but at the same time criticizes contemporary geopolitics. He highlights the fact that the Indian Ocean welcomes sailors from India, China, Mauritius, Europe, and the United States, but the trade was colonial with the terms being dictated by the British Empire. Even today, the ocean remains the lieu of maritime trade affecting politics on land, agricultural production, and environmental policy making. This archive of unfair trade is evident in the IOR-ARC treaty signed by countries that share the Indian Ocean. Though created with the intention of being a platform for the peoples of the Indian Ocean Region to reconnect with each other, to discover their common heritage and deep-rooted affinities, to celebrate their shared cultural history, and to chart their own destinies, the free trade association has been criticized for having pitched "too high" or "too low" its tariffs and customs barriers. In India, the cultivation of opium poppies is now regulated with farmers producing mainly for medicinal or research purposes. Governmental proposals to create a drug-free state at the expense of poor farmers has resulted in these farmers losing their livelihood. He insists that mere ecofriendly policy labeling will not solve the world's problems. He calls on writers to recall and to write on issues that affect the world and to consider new econarratives that speak of the need to preserve life, encourage environmental advocacy, and bring ecocriticism closer to the material turn by highlighting how narratives and stories contribute to making meaning of the material forces and substance that rule the world. Ghosh's ecocritical writing shows that ignoring climate change and environmental hazards will certainly make future generations contemplate why their predecessors encouraged this time of Great Derangement.

Roy refers to the disparity between the rich and the poor as "this confederation of loyal, corrupt, authoritarian governments in poorer countries to push through unpopular reforms and quell mutinies."[103] Through an enumeration of the Tata empire's material influence on our daily life, Roy contends that the ordinary man is "under siege": "We all watch Tata Sky, we surf the net with Tata Photon, we ride in Tata taxis, we stay in Tata Hotels, sip our Tata tea in Tata bone china and stir it with teaspoons made of Tata Steel. We buy Tata books in Tata bookshops. *Hum Tata ka namak khatey hain.*[104] We're under siege."[105] The essay-manifesto is the ideal mode for Roy's ecocritical enterprise of underlining the need for social and political change. It speaks of Roy's envisioning of a "postcolonial green" that campaigns for the transference from "red" to "green" politics and the need to take the material turn and dwell as responsible inhabitants who believe in global justice and sustainability on our planet Earth. Roy posits that the world needs a "new kind of politics. Not the politics of governance, but the politics of resistance. . . . The politics of forcing accountability."[106] She advocates writing that contributes to climate education and encourages the youth to join hands across the world and prevent the destruction of the planet Earth.

Amitav Ghosh advances the idea that a writer's imagination plays a vital role in shaping the minds of young readers. He states that "Fiction, for one, comes to be reimagined in such a way that it becomes a form of bearing witness, of testifying, and of charting the career of the conscience."[107] Through her writing, a form of nonviolent dissent to change the world, Arundhati Roy rightly advocates that "the only thing worth globalizing is dissent. It's India's best export."[108] The pedagogical approach of analyzing the climate change and sustainability challenges of the century through the lens of making a material turn in fiction and nonfiction is the ideal mode to capture the attention and interest of the youth at large.

NOTES

1. This chapter was previously published as Suhasini Vincent, "Amitav Ghosh and Arundhati Roy on Climate Change: A Pedagogical Approach to Awakening Student Engagement in Ecocriticism," in *Literature as a Lens for Climate Change: Using Narratives to Prepare the Next Generation*, ed. Rebecca Young (Lanham: Lexington Books, 2022), 53–76.

2. Amitav Ghosh has a doctorate in social anthropology from the University of Oxford. His works reveal his study of the ways in which people live in different social and cultural settings across the globe. His other works of fiction are *The Circle of Reason* (1986), *The Shadow Lines* (1988), *In an Antique Land* (1992), *The Calcutta Chromosome: A Novel of Fevers, Delirium, and Discovery* (1995), *The Glass Palace* (2000), *The Hungry Tide* (2004), the Ibis Trilogy comprising *Sea of Poppies* (2009), *River of Smoke* (2011), *Flood of Fire* (2015), and *Gun Island* (2019).

3. Arundhati Roy (born November 24, 1961) is an Indian author, actress, and political activist-writer who was awarded the Booker Prize in 1997 for her semi-autobiographical debut novel *The God of Small Things*. This essay will highlight her involvement in environmental and human rights causes.

4. Ghosh's other works of nonfiction include *Dancing in Cambodia, at Large in Burma* (1998), *The Imam and the Indian* (2002), *and Incendiary Circumstances: A Chronicle of the Turmoil of Our Times* (2005).

5. Amitav Ghosh, *The Great Derangement: Climate Change and the Unthinkable* (Chicago Illinois: The University of Chicago Press, 2016), 10.

6. Ghosh, *The Great Derangement: Climate Change and the Unthinkable*, 11.

7. Ghosh, 11.

8. Arundhati Roy's political essays have been compiled into the following collections: *The End of Imagination* (1998), *The Algebra of Infinite Justice* (2001), *Power Politics* (2002), *War Talk* (2003), *An Ordinary Person's Guide to the Empire* (2005), *The Shape of the Beast* (2008), *Listening to Grasshoppers: Field Notes on Democracy* (2009), *Broken Republic* (2011), and *My Seditious Heart* (2019).

9. In a legal affidavit titled "On Citizens' Rights to Express Dissent" in her work *My Seditious Heart*, Roy quotes an extract from the Supreme Court Order dated October 15, 1999, that accuses Roy's writing of violating the dignity of the court and polluting the stream of justice.

10. Arundhati Roy, "On Citizens' Rights to Express Dissent," in *My Seditious Heart: Collected Nonfiction* (Toronto: Hamish Hamilton, 2019), 156.

11. Arundhati Roy has received awards for her essay writing, like the Lannan Award for Cultural Freedom in 2002, the 2003 Noam Chomsky Award, the Sydney Peace Prize in 2004, the 2005 Sahitya Kademi Award, which she declined, and the Norman Mailer Prize in 2009 for Distinguished Writing.

12. Serenella Iovino and Serpil Oppermann, *Material Ecocriticism* (Indiana University Press, 2014), 1–2.

13. Ghosh, *The Great Derangement: Climate Change and the Unthinkable*, 8.

14. Ghosh, 9.

15. Arundhati Roy, "Arthur Miller Freedom to Write Lecture," *The Guardian*, May 13, 2019, par. 2–5. https://www.theguardian.com/commentisfree/2019/may/13/arundhati-roy-literature-shelter-pen-america.

16. Roy, par. 8.

17. Roy, par. 8.

18. Ghosh holds a PhD in social anthropology from the University of Oxford. He also received honorary doctorates from Queen's College in New York and the Sorbonne.

19. Roy has a degree in architecture from the Delhi School of Architecture. She is also a screenplay writer and has written the screenplays for *In Which Annie Gives It Those Ones* (1989), a film based on her experience as a student of architecture.

20. Roy criticizes the dominance of richer nations over developing countries by means of unequal conditions of economic exchange. She claims that neoimperialism results in richer nations restricting poorer ones from stepping out of the roles that the former had defined for them, like for instance reducing the latter to providers of raw materials and cheap labor.

21. https://www.amitavghosh.com/; https://www.weroy.org/

22. Ghosh, *The Great Derangement: Climate Change and the Unthinkable*, 15–17.

23. The meandering River Padma is one of the major rivers of Bangladesh. Evidence from satellite imagery reveal that it has been constantly gaining volume and changing its trajectory during the last decade. The river's source is at the junction of the Ganges and Jamuna Rivers in India. It then merges with the Meghna River in Bangladesh and ultimately empties into the Bay of Bengal.

24. The Ganges River (Ganga in Hindi) is the great river of the plains of northern Indian sub-continent. It rises in the Himalayas and empties into the Bay of Bengal. As it flows through the Indo-Gangetic Plain, it irrigates one-fourth of the country, and has

been the cradle of successive civilizations from the Mauryan empire of Ashoka in the third century BC to the Mughal Empire, founded in the sixteenth century.

25. In *The Hungry Tide*, Ghosh speculates on the origin of the name of the mangrove forests and the anthropological, botanical, geotidal, and historical influences in the bearing of its name. "Sundarbans" means "beautiful forest."

26. Ghosh, *The Great Derangement: Climate Change and the Unthinkable*, 6.

27. Bhatir Desh means "land of the low tide" or "tidal country." Ghosh explains "in the record books of the Mughal emperors this region is named not in reference to a tree but to a tide—bhati. And to the inhabitants of the islands this land is known as bhatir desh—the tide country—except that bhati is not just the 'tide' but one tide in particular, the ebbtide, the bhata: it is only in falling that the water gives birth to the forest. To look upon this strange parturition, midwived by the moon, is to know why the name 'tide country' is not just right but necessary" (Ghosh 2004, 8).

28. Amitav Ghosh, *The Hungry Tide* (New Delhi: Ravi Dayal, 2004), 7.

29. Hugli River, also spelled Hooghly, is a river in West Bengal state, in northeastern India. It is an arm of the Ganges (Ganga) River. It branches off the Ganges and provides access to Kolkata (Calcutta) from the Bay of Bengal.

30. The Meghna River is the major watercourse of the Ganges in Bangladesh. It receives the combined waters of the Padma and Jamuna (the name of the Brahmaputra in Bangladesh) rivers near Chandpur in Bangladesh.

31. Ghosh, *The Hungry Tide*, 246.

32. Bon Bibi or Ban Bibi is a reference to the guardian spirit of the Sundarbans who protects woodcutters, inhabitants, and travelers from tiger attacks.

33. Dokhin Rai or Dakshin Rai is a revered deity in the Sundarbans who rules over beasts and demons. He is the arch enemy of Bon Bibi.

34. After a twenty-year gap of novelistic silence, Roy creates a fictional space in *The Ministry of Utmost Happiness* that reveals the lives of hijras (eunuchs) who live in communities that have created alternative structures of kinship, resistance, and romance.

35. Arundhati Roy, *The Ministry of Utmost Happiness* (New Delhi: Penguin, 2017), 195.

36. Roy Arundhati, "Power Politics," in *The Algebra of Infinite Justice* (New Delhi: Penguin, 2002), 147.

37. Arundhati Roy, *Capitalism: A Ghost Story* (London: Verso, 2015), 13.

38. Roy, *Capitalism: A Ghost Story*, 2.

39. Ironically Vedanta's tagline is "Mining Happiness." Its British-based parent company Vedanta Resources, a natural resources conglomerate, aims at creating long-term shareholder value through research, discovery, acquisition, sustainable development, and utilization of natural resources. The company mines natural resources, but Roy muses on whether they really mine happiness.

40. In 2014, the Jindal Group Jindal was under investigation in connection with the allocation of coal mining rights in a scandal called "Coalgate" by India's media.

41. The Pollution Control Board of the State of Madhya Pradesh has imposed a fine of fifty lakh rupees (approximately $70,275) towards farmers' compensation. The

Essar plant was responsible for crop damage caused by overflowing ash water from one of its power plants in Singrauli.

42. Think Festivals were organized during the period 2011–2013 and aimed at creating thought-provoking, egalitarian ideas from across the globe.

43. It was begun in 1907 by Jamshedji Tata, India's pioneer industrialist. It is today one of the world's most geographically diversified steel producers with operations and commercial presence across the world.

44. The Anglo-Australian multinational is engaged in diamond mining in the Indian state of Madhya Pradesh.

45. JLF is an annual literary festival held in the Indian city of Jaipur. It is a five-day festival where writers, thinkers, politicians, and entertainers meet on one stage to engage in thoughtful debate and dialogue.

46. Salman Rushdie, "Is Nothing Sacred?," in *Imaginary Homelands: Essays and Criticism, 1981–1991* (London: Vintage Books, 2010), 428.

47. In his essay "Is Nothing Sacred?" Salman Rushdie leads the reader into the unassuming little room of literature tucked away in a corner of the large rambling house of world activity. The room is alive with ideas and dialogues. Rushdie articulates the dire need to preserve this privileged arena of creative enterprise and includes readers in the enterprise of assuring the survival of the little unassuming room of literature by putting forth the idea that creating a literary masterpiece is not an affair of individual creative genius, but a joint endeavor.

48. Roy, *Capitalism: A Ghost Story*, 19.

49. An acronym coined by Jim O'Neill of Goldman Sachs in 2001. The BRIC countries comprising of Brazil, Russia, India, and China are not a political alliance like the European Union or a formal trading association. They assert their power as an economic bloc by signing formal trade agreements together and attending summits together. In 2010, South Africa joined this group and they now concert with each other's interests as BRICS countries. By 2050, these countries will probably be wealthier than most of the current economic powers.

50. In the 1950s, the name "the Golden Triangle" was given to the area of mainland Southeast Asia where most of the world's illicit opium originated. This name "Golden Triangle" was first coined by the United States vice secretary of state, Marshall Green, during a press conference on July 12, 1971.

51. Pierre-Arnaud Chouvy, *Opium: Uncovering the Politics of the Poppy* (Cambridge, MA: Harvard University Press, 2010), 23.

52. Amitav Ghosh, *Sea of Poppies* (London: John Murray, 2008), 12.

53. The Andaman Sea takes its name from the Andaman Islands, a union territory of India. It is situated in the northeastern Indian Ocean. It is the sea link between Myanmar and other South Asian countries. It is on the shipping route between India and China, via the Strait of Malacca.

54. A reference to the old foreign enclave in Canton (Guangzhou) which was also known as the "Thirteen Factories" (Saap Sam Hong). Fanqui-town does not exist today as these factories were burned down in 1856 and were never rebuilt again.

55. The novel's diegesis reveals how the British imposed their trade in Indian opium upon China through the two so-called Opium Wars (1839–1842 and

1856–1860), thus not only monopolizing the delivery of the narcotic to the European market, but also changing the course of economic and political relations between the East and the West.

56. Carl A. Trocki, *Opium, Empire, and the Global Political Economy: a Study of the Asian Opium Trade 1750–1950* (London: Routledge, 1999), 7–8.

57. Julie Rivkin, Michael Ryan, and Edward Said, "Orientalism," in *Literary Theory, an Anthology* (Malden, MA: Blackwell, 1998), 881.

58. Edward Said in his work Orientalism describes "Orientalism" as the Western attitude that views Eastern societies as exotic, primitive, and inferior.

59. It is interesting to make a reference to *Opium: Uncovering the Politics of the Poppy*, where Pierre-Arnaud Chouvy explains how a triangular trade developed between Britain, India, and China in which Indian opium provided the silver required to buy tea legally from China for shipments to London. He speculates on how opium ensured the profitability of colonial trade by enabling the British to balance their trade deficit with China: "Fearing that payment for Chinese imported goods (tea consumption was growing fast in Britain when only China produced the leaves) would deplete their silver reserves, the British resorted to opium, a product of their Indian colony, as a means of payment." It is interesting to know that through the opium trade of the English East India Company and the tax on Malwa opium, the British made a profit of $15,488,000.

60. *Nemesis* is the first British ocean-going iron warship. It was launched in 1839 by the British East India Company to take part in the Opium Wars. It was also known as "The Devil Ship" due to the havoc it stirred up during attacks.

61. Amitav Ghosh, *Flood of Fire* (London: John Murray, 2016), 509–10.

62. Ghosh, 505.

63. The phrase "The Empire Writes Back to the Centre" was originally used by Salman Rushdie in his article "The Empire Writes Back with Vengeance." Rushdie's pun is also from the film *Star Wars: The Empire Strikes Back*. In his article, Rushdie referred to the erstwhile British colonies as the Empire as they regained independence and wrote back to the former colonizer. Many diasporic writers in the United Kingdom have used various literary strategies of decolonization to set the record straight by writing back to the Centre or the former British Raj. Australian critics, Bill Ashcroft, Gareth Griffiths, and Helen Tiffin titled their work *The Empire Writes Back* to highlight the powerful forces acting on language in the postcolonial text and show how these texts constitute a radical critique of Eurocentric notions of literature and language.

64. Bill Ashcroft, Gareth Griffiths, and Helen Tiffin, *The Empire Writes Back: Theory and Practice in Post-Colonial Literatures* (Brantford, ON: W. Ross MacDonald School Resource Services, 2011), 221.

65. Ghosh, *Flood of Fire*, 484.

66. The area of Southeast Asia encompassing parts of Burma, Laos, and Thailand, significant as a major source of opium and heroin.

67. The name "Golden Crescent" is of unknown authorship and is similar to that of "Golden Triangle" and can be considered as the Southeast Asian alter ego. "Its 'Crescent' refers to the Muslim dimension of this opium-producing region comprising the

countries of Afghanistan, Iran, and Pakistan . . . Afghan and Pakistani opium was first marketed for Persian and Indian consumers" (Chouvy 28–29). So, the name "Golden" refers to the Muslim dimension of this opium-producing region comprising the afore-mentioned countries. Interestingly, Afghan and Pakistani opium was first marketed for Persian and Indian consumers.

68. Roy, *Capitalism: A Ghost Story*, 8.

69. Roy, 7.

70. Roy, 9.

71. Roy, 9.

72. Roy, 9.

73. Roy, 9.

74. Roy, "Power Politics," 150–151.

75. Roy, 152.

76. Three mineral-rich Indian states; mineral-rich Orissa where hundreds of indig-enous tribespeople are battling to stop London-listed Vedanta Resources Plc from extracting bauxite from what they say is their sacred mountain.

77. Roy, *Capitalism: a Ghost Story*, 11–12.

78. A Latin term from the history of pharmacy. It is the branch of medical science that deals with the sources, nature of plants, properties, and preparation of drugs.

79. The dried young leaves of camellia sinensis have been used since ancient times to make green tea. It contains a high concentration of antioxidants known as polyphe-nols as the leaves are not fermented.

80. Amitav Ghosh, *River of Smoke* (London: John Murray, 2011), 39.

81. Ghosh, 82–83.

82. The green paradigm refers to the quest of embracing social and ecological values that protect the planet.

83. Roy, *Capitalism: A Ghost Story*, 15.

84. The building of Kalpasar dam in Gujarat is still underway. Once completed, it will be thirty-four kilometers long stretching across the Gulf of Khambat. There are plans to configure a ten-lane highway and a railway line on top of it. This sweet water reservoir of Gujarat's rivers has a further confluence and network of 168 dams that are mostly privatized and have been planned across war-hit and tension-prone zones of Kashmir and Manipur.

85. Special Investment Region (SIR) refers to an investment region of 100 sq. kms in Gujarat region of India. The aim is to set up world class hubs of economic activity. SEZ refers to Special Economic Zones where government agencies including private companies may be assigned powers and functions to promote the development of a Special Investment Region.

86. Roy, *Capitalism: A Ghost Story*, 16.

87. The Adivasis are the earliest inhabitants of the Indian subcontinent, who are considered as the indigenous tribes who occupied the hill and mountainous regions of India. The term Adivasi was coined in the 1930s to give a sense of identity to the various indigenous tribes of India. In Hindi *adi* means "of earliest times," and *vasi* means "inhabitant."

88. The Sarvar Sarovar Project across the Narmada is being built at an estimated cost of Rs. 392.4 billion (approximately eight billion USD). It is the world's second largest concrete gravity dam, with the world's third highest spillway discharging capacity of 87,000 m³ and will be the largest irrigation canal in the world after its completion, irrigating Kutch and Saurashtra. The government of India calls it an "ecofriendly" indigenous hydropower reservoir.

89. Roy Arundhati, "Greater Common Good," in *The Algebra of Infinite Justice* (New Delhi: Penguin, 2002), 46.

90. The river Narmada is the fifth longest river in the Indian subcontinent and flows through central India, through the state of Madhya Pradesh. Thirty mega-dams are being constructed along the river basin, which is home to teak forests.

91. Roy, "Greater Common Good," 47.

92. The ironic reference to suffering for the cause of common good mentioned in Nehru's "dam" speech has inspired the title of the essay "Greater Common Good."

93. The Hirakud dam is India's first river valley project after India's independence (August 14, 1947). It is India's largest dam and the world's fourth largest barrage. It provides hydroelectricity to villages along the Mahanadi River.

94. Roy, "Greater Common Good," 69.

95. Roy, 52.

96. Roy, "Power Politics," 155.

97. The Dholera Special Investment Region (DSIR) covers approximately 920 sq km and covers twenty-two villages that are strategically situated between the industrial zones of Ahmedabad, Baroda, Rajkot, and Bhavnagar in Gujarat State. This industrial hub will have a six-lane-access-controlled expressway, metro rails, and an international airport.

98. Graham Huggan, *Postcolonial Ecocriticism: Literature, Animals, Environment* (London: Routledge, 2015), 5.

99. Graham Huggan and Helen Tiffin in *Postcolonial Ecocriticism* examine the relationship among humans, animals, and the environment in postcolonial literary texts. They hold that human societies need to consider their relationship with nonhuman species with whom they share the planet. They insist on the need to imagine new ways of creating awareness of these ecologically connected groupings.

100. Huggan, *Postcolonial Ecocriticism*, 12

101. Roy, "Greater Common Good," 109.

102. This echoes Rushdie's preference for an aesthetics of inconstancy in writing. He claims: "Stasis, the dream of eternity, of a fixed order in human affairs, is the favoured myth of tyrants; metamorphosis, the knowledge that nothing holds its form, is the driving force of art" (Imaginary 291).

103. Arundhati Roy, "Confronting Empire," in *My Seditious Heart: Collected Nonfiction* (Toronto: Hamish Hamilton, 2019), 223.

104. Meaning "we eat Tata salt" in Hindi.

105. Roy, *Capitalism: A Ghost Story*, 20

106. Roy Arundhati, "The Ladies Have Feelings," in *The Algebra of Infinite Justice* (London: Flamingo, 2002), 215.

107. Ghosh, *The Great Derangement: Climate Change and the Unthinkable*, 128.

108. Roy "The Ladies Have Feelings," 215.

BIBLIOGRAPHY

Ashcroft, Bill, Gareth Griffiths, and Helen Tiffin. *The Empire Writes Back: Theory and Practice in Post-Colonial Literatures*. Brantford, ON: W. Ross MacDonald School Resource Services, 2011.

Chouvy, Pierre-Arnaud. *Opium: Uncovering the Politics of the Poppy*. Cambridge, MA: Harvard University Press, 2010.

Ghosh, Amitav. *The Hungry Tide*. New Delhi: Ravi Dayal, 2004.

———. *Sea of Poppies*. London: John Murray, 2008.

———. *River of Smoke*. London: John Murray, 2011.

———. *Flood of Fire*. London: John Murray, 2016.

———. *The Great Derangement: Climate Change and the Unthinkable*. Chicago Illinois: The University of Chicago Press, 2016.

Huggan, Graham. *Postcolonial Ecocriticism: Literature, Animals, Environment*. London: Routledge, 2015.

Iovino, Serenella, and Serpil Oppermann. *Material Ecocriticism*. Indiana University Press, 2014.

Rivkin, Julie, Michael Ryan, and Edward Said. "Orientalism." Essay. In *Literary Theory, an Anthology*, 873–85. Malden, MA: Blackwell, 1998.

Roy, Arundhati. "Greater Common Good." In *The Algebra of Infinite Justice*, 46–141. New Delhi: Penguin, 2002.

———. "Power Politics." In *The Algebra of Infinite Justice*, 143–84. New Delhi: Penguin, 2002.

———. "The Ladies Have Feelings." In *The Algebra of Infinite Justice*, 187–215. London: Flamingo, 2002.

———. *Capitalism: a Ghost Story*. London: Verso, 2015.

———. *The Ministry of Utmost Happiness*. New Delhi: Penguin, 2017.

———. "Arthur Miller Freedom to Write Lecture." *The Guardian*, May 13, 2019, https://www.theguardian.com/commentisfree/2019/may/13/arundhati-roy -literature-shelter-pen-america.

———. "Confronting Empire." Essay. In *My Seditious Heart: Collected Nonfiction*, 221–26. Toronto: Hamish Hamilton, 2019.

———. "On Citizens' Rights to Express Dissent." Essay. In *My Seditious Heart: Collected Nonfiction*, 150–59. Toronto: Hamish Hamilton, 2019.

Rushdie, Salman. "Is Nothing Sacred?" Essay. In *Imaginary Homelands: Essays and Criticism, 1981–91*, 415–29. London: Vintage Books, 2010.

Trocki, Carl A. *Opium, Empire, and the Global Political Economy: a Study of the Asian Opium Trade 1750–1950*. London: Routledge, 1999.

Vincent, Suhasini. "Amitav Ghosh and Arundhati Roy on Climate Change: A Pedagogical Approach to Awakening Student Engagement in Ecocriticism." In *Literature as a Lens for Climate Change: Using Narratives to Prepare the Next Generation*, edited by Rebecca Young, 53–76. Lanham: Lexington Books, 2022.

Chapter Ten

Creating Authentic Learning Experiences

Interdisciplinary Climate Change Instruction and Assessment

Mary-Alice Corliss and Rebecca L. Young

"As educators representing schools from around the world, with expertise in a range of disciplines across grade levels, how can we respond to the now global youth movement's call to action for comprehensive climate education? Students are asking for it, and we can deliver."

The conference room stilled, discomfort consuming a space that moments before had been filled with the energy of academics eager to share ideas about the future of education. It was late 2019, school strikes had been taking place for over a year as young activists were gaining momentum worldwide in communicating, organizing, and protesting. With colorful homemade placards, banners, T-shirts, blogs, and social media posts, strikers demanded immediate action on climate change—from education reform to policy action. It seemed appropriate to ask this group of leaders, who had gathered for a conference on international education and assessment, to consider how we might respond. Instead, deafening silence was quickly followed by all the reasons why schools should not be asked to take up this effort—it is too political, too controversial, too massive.

When students rally on this scale—with marches and walk-outs now having taken place in over 150 countries—to demand we prepare them for as monumental a challenge as climate change, we ought to take them seriously. What it means to work cooperatively, to problem-solve across disciplines, to communicate information for all stakeholders, is constantly evolving as climate scientists unabashedly raise the alarm for immediate action. Governing bodies seek innovative solutions to increasingly complex data-informed scenarios that impact our present and future. How graduates will manage the formidable challenges ahead, let alone the many daily decisions they make as consumers in a climate-changed world, should be an essential question driving the pedagogy of every academic community around the globe.

Twenty-first-century learner skills can be found thoughtfully expressed in the mission statements of educational institutions around the world, but the extent to which we are meaningfully fostering and assessing students' ability to problem-solve in real-world contexts remains limited or more typically isolated at most K–12 grade levels; the structure of higher education schedules makes this similarly difficult which is even more concerning as students are nearer to attaining professional degrees. These factors reduce the likelihood that individual classroom educators will integrate realistic independent or collaborative contexts into their required content. Determining which specific soft skill capabilities should be prioritized is a subject of ongoing debate that further complicates this effort. As a point of reference for this chapter, the National Resource Council (NRC), in its introduction to *Assessing 21st Century Learning Skills*, summarizes that these skills traditionally include

> being able to solve complex problems, to think critically about tasks, to effectively communicate with people from a variety of different cultures and using a variety of different techniques, to work in collaboration with others, to adapt to rapidly changing environments and conditions for performing tasks, to effectively manage one's work, and to acquire new skills and information on one's own.[1]

As part of a broader workshop to explore how changes in the workplace are determining the shift in what skills employers seek, NRC authors noted that "five skills appear to be increasingly valuable: adaptability, complex communication skills, non-routine problem-solving skills, self-management/self-development; and systems thinking."[2]

Pursuing this inquiry to its source, Harvard School of Education economist Richard Murnane shared findings of workplace dynamics that illuminate this growing demand for soft skills, specifically those that require "expert thinking" and "complex communication," over those that are considered "routine manual" and "routine cognitive" skills. As helpfully summarized in the NRC's report, Murnane's additional research in these two areas highlight key workplace-related expectations:

- Within a domain, workers need a deep understanding of the domain and relationships within it.

- Pattern recognition
- A sense of initiative (i.e., when you see a new task, is this a challenge you are anxious to take on or one you shy away from?)
- Metacognition (i.e., monitoring your own problem solving)

Likewise, the components of complex communication include the following:

- Observing and listening
- Eliciting critical information
- Interpreting the information
- Conveying the interpretation to others[3]

Murnane's conclusions about the importance of these skill sets as "essential for 'leading a contributing life in a pluralistic democracy'" are particularly significant as we consider them in the context of a changing world. As the NRC authors summarize:

> He enumerated the complex set of problems that the [United States] faces, including such issues as immigration, global warming, and proliferation of nuclear weapons. In his view, understanding these problems and participating in their solutions requires a well-educated citizenry adept at expert thinking and complex communication.[4]

Indeed, as this chapter will explore, the importance of these skills goes far beyond workplace demands to encompass what it means to be a productive, responsible citizen, especially in a democracy. Crucially, this includes what it means to be a thoughtful steward of this shared planet and its natural resources—what it means to be a conscientious global citizen.

As Tony Wagner urges in *Creating Innovators: The Making of Young People Who Will Change the World*, "what you know is far less important than what you can do with what you know. The interest in and ability to create new knowledge to solve new problems is the single most important skill that all students must master today."[5] Climate change is perhaps today's most relevant—certainly its most urgent—lens through which to teach, foster, and assess the skills of innovators, which Wagner and others similarly outline as "collaboration; multidisciplinary learning; thoughtful risk-taking, trial and error; creating; intrinsic motivation: play, passion, and purpose."[6] To Murnane's point about the importance of a well-educated citizenry in confronting extremely complex geopolitical issues, the need for leaders in education to act in support of such skills is critical: it is through education that we confront these interrelated problems.

Educational assessments that evaluate students' informed readiness to seek solutions in proactive, prosocial ways can provide valuable feedback about soft skill competencies that will impact all areas of a student's life—from home to workplace relationships to larger social or civic engagements. Such formative and summative assessments also inform continuous improvement for curricular objectives, pedagogical practices, and school-wide goals related to twenty-first-century learning. Ecopedagogy enhanced by climate contexts allows educators to evaluate not only content-based competencies but skills like collaborative cooperation, interdisciplinary problem-solving, creativity, and communication. Acknowledging that full-scale curricular reform which includes climate education across grade levels and subject areas is the imperative, climate-focused authentic assessment can help align curricular objectives to the urgency of action this crisis demands.

As the chapters in this book illustrate, a framing lens that addresses environmental concerns in the context of a story can be an effective way to introduce or delve into climate education. In this chapter, we explore an assessment approach that pairs literature with science to help students understand the complexity of a problem by imagining its interrelated causes and solutions.

Coupling scientific data with narrative creates a foundation upon which a teacher can check student understanding of climate change-related phenomena through various instructional and summative assessment tasks. Because this chapter concerns developing assessments as part of climate change instruction, the focus will be formative evaluations of learning. Further, it should be noted that just as a teacher has the flexibility to choose a particular narrative, how that narrative will be presented, and what additional information and data will supplement that narrative, a teacher can use this approach to target a variety of science domains and standards in assessment. This includes the opportunity to incorporate twenty-first-century skills (e.g., collaboration, problem-solving) and even social-emotional learning (empathy, interpersonal and intrapersonal awareness and skills), with the awareness that because social-emotional learning (SEL) competency develops over time this should be reflected in assessment practices.[7]

For example, a high school life science teacher could frame a series of lessons or even an entire ecosystem unit around a narrative that explores the topic of coral reef depletion due to ocean acidification caused by climate change. Documentaries like *Chasing Coral* and even children's stories such as *Zobi and the Zoox: A Story of Coral Bleaching* can support this narrative focus with explanations of the science, while a fictional lens might foster a more nuanced understanding of the plight of coral reefs. One example is Romesh Gunesekera's *Reef*, which explores the political and socioeconomic contexts that drive the destruction of coral reefs in Sri Lanka. Triton, the novel's narrator, recalls the time he spent as a child with a marine biologist desperate to preserve remaining coral reefs, reflecting on the conflict for local coral minors and fishermen who depend on the coastal ecosystem they are systematically depleting:

> Mister Salgado only slowed down when we came to the skull-heaps of petrified coral—five-foot pyramids beside smoky kilns—marking the allotments of a line of impoverished lime-makers, tomorrow's cement fodder, crumbling on the loveliest stretch of the coast.[8]

Though a global problem, this setting's descriptions connecting the human-driven destruction of Sri Lanka's coastal beauty with the desperate consequences it will have on local communities make this narrative a compelling study for understanding the concept of disproportionate impact.

It is important to note here that there is a significant difference between the teacher initially choosing a relevant topic and then leveraging that topic to help students make sense of the associated phenomenon.[9] A phenomenon is more than simply a topic: it is an observation that requires explanation or a problem that requires a solution. When framed in a storyline, students can use that context to develop a deeper understanding of the phenomenon. To take this topic and home in on a related phenomenon, the teacher could indicate that corals are becoming bleached more often now than in the past, which sets up an intriguing mystery for students to explore. From there, the teacher could contextualize this observation by selecting excerpts from the narrative that highlight perspectives of peoples affected by this phenomenon, associating descriptions of what occurs in the story to actual data on past and present bleaching events, including scientific predictions of further impacts. The science content addressed could focus on ecosystem standards, inviting students to reflect on the long-term consequences of climate change upon ecosystems and humanity. Then, they could be required to justify their understanding with evidence. Another option would be to evaluate technologies currently used to address bleaching and ocean acidification. A fourth avenue for investigation might be to determine possible solutions for limiting the impact of these events on coral reefs. This is one example of a topic ripe for teaching the science of climate change; if the appropriate narrative and learning targets are selected, the approach would be valid in any K–12 grade level or higher education course. The inclusion of a narrative allows teachers to assess literary skills, but more imperatively students' interpretation of characterizations encourages them to take on the perspectives of peoples who are disproportionately affected by the impacts of climate change. This process would naturally raise discussions of equity within the scale of this phenom-enon as an inherent part of the narrative's storyline. An SEL component may include asking students to evaluate if and how their level of concern for the characters' real-life counterparts was impacted by the story and their explora-tion of the science.

A variety of formative assessment classroom techniques (FACTs) can be used as part of this instructional approach to support deeper learning of climate change. Pre-instruction strategies can be used prior to students read-ing a narrative so that the teacher can identify instructional goals that target misunderstandings and misinformation.[10] During instruction, collaborative, argument-based formative tasks would then work very well, since an emo-tionally engaging storyline with character perspectives and connections to actual data would prompt a humanity-centered focus on content. Using these FACTs would be useful if a teacher wanted to assess SEL interpersonal skills such as social perspective-taking and collaborative problem-solving, which follow from empathy (interpersonal awareness). Especially for climate-change phenomena, assessments that focus on how students develop a sense of self-efficacy (intrapersonal awareness) and stress management (intraper-sonal skill) in response to a crisis could also be conducted in this setting[11] regardless of whether data is collected in an in-person or virtual learning environment.

This approach could also inform the use of claims-evidence-reasoning assessments by requiring students to synthesize various kinds of evidence (both from scientific sources and through literature citation) to support claims and connect that evidence with reasoning. This can be especially use-ful for assessments that are constructed using evidence-centered design,[12] as the teacher could evaluate how effectively students relate information in a narrative to actual data. Teaching climate-change science using a narrative can therefore be effective for use with a variety of assessment strategies, instructional objectives, and pedagogical practices. The emphasis of this research-based approach is to promote deeper learning of the science, with a narrative lens that supports twenty-first-century skill development and SEL. Future generations of learners who understand the complexities of climate change will have the skills to address crises that result from it.

Well-designed instructional and assessment tools that use literature as a lens for science could be potentially transformational in students' understand-ing of climate change. The rationale behind this approach leverages cogni-tive science research on theory of mind (ToM), which refers to "our ability to explain people's behavior in terms of their thoughts, feelings, beliefs, and desires."[13] As novelist and cognitive psychologist Keith Oatley explains, the idea of Theory of Mind refers to knowing what another person may be thinking or feeling in the moment, but with people whom we are likely to interact with on future occasions, we gather information to build mental models of them over the long term, from their utterances, from their behavior, and from what other people say about them.[14]

Since we naturally employ this knowledge in our reading lives, we can use it to understand situations and points of view that may be different from our own. A climate-related scientific phenomenon contextualized by narrative engages our empathy toward characters and conflicts, thereby helping readers respond to the attitudes and information presented. As noted in *Assessing 21st-Century Skills*, "successful interpersonal behavior involves a continuous correction of social performance based on the reactions of others . . . a form of 'social intelligence,' . . . social perception and social cognition . . . involve processes such as attention and decoding."[15] The very act of reading appears to be potentially excellent practice for self-improvement in social interactions because it hones our predictive and interpretive skills. Recommendations for studying climate change narratives are thus informed by students' intuitive social intelligence or ToM capabilities.[16]

Recognizing that an excerpted storyline paired with scientific data may be students' first interaction with interdisciplinary assessment, their interpretation of characters' states of mind and emotions (whether they intuit them correctly or incorrectly or feel similarly or not) will prepare them for engaging with supplemental stimuli (e.g., data sets, science simulations) selected to inform climate-related concerns highlighted by the fiction. By fostering students' appreciation for how they can participate in affecting positive change in any number of real-world contexts, this approach endeavors to channel the power of emotional awareness established through the narrative into student agency and prosocial action.

The cognitive constructivism roots of authentic learning emphasize individual sense-making, so that learning is experienced through a sequence of mental processes—starting with the retrieval of prior knowledge, then a "hook" that introduces new information and/or disrupts prior knowledge, application of knowledge, and, finally, reflection.[17] While this sequence is more familiar in an instructional setting, the same processes can, and should, be mirrored in the framing and assessment of science phenomena because they are compatible with how the human brain retrieves and applies information to problem-solve. Climate change is a complex topic that requires various large-scale solutions to effectively address its many impacts. Offsetting these impacts requires education that is focused on interdisciplinary comprehension and prosocial innovation, making reliable assessments of twenty-first-century skills essential. Individuals in every community role and in all workplaces need to understand the problem before they can contribute to solutions.

An important aspect of a constructivist approach to science education is the "hook" between old and new information, because it creates the bridge between retrieved memories and new understandings of those memories. This "hook" can simply be an interesting piece of information, something that contradicts prior knowledge, or, more significantly, an emotional connection that makes the information local or personal. If this emotional connection is not so distracting that it sidetracks students, it becomes a powerful tool for promoting learning and eliciting understanding. A story that frames an aspect of climate change can be very useful in the classroom because it involves a "hook" to bridge new and old information, as well as fictional experiences to actual data and phenomena—the teacher has the flexibility to select parts of a narrative that target key concepts, promote critical thinking, and involve students in perspective-taking.

Assessment stimuli, if chosen carefully, can help students appreciate how the many phenomena climate change encompasses affect their lives and the lives of others. This can be a useful teaching tool, as active emotional engagement is effective for learning, and emotional memory helps stimulate other memory pathways for retrieval.[18] Assessment contexts that have emotional meaning for learners are therefore not only more compelling but also useful for eliciting understanding—an important aspect of any well-designed assessment.

For students to not only be emotionally invested but feel comfortable enough to acknowledge the realities of climate change, selection and presentation of the climate change narrative need to be determined with specific psychological concepts in mind as they relate to crisis communication. For example, the storyline should make the crisis personal, rather than merely exploiting it as a dystopian backdrop, to help readers process complex information. Narrative strategies that pull readers in to support a particular character or to root for a certain outcome are thus helpful in making this selection.

Since climate change impacts certain communities around the world more directly and immediately than others, it is challenging to get some learners emotionally engaged in learning about and addressing its influence on their own. Data and case studies of long-term and short-term impacts are one way to make learners aware, but for a learner to empathize with distant communities and recognize how their own actions affect such communities, it is important to help students take on another perspective. However, presenting narratives of a large-scale crisis without providing a clear storyline or potential solutions has been shown to cause people to become indifferent to suffering,[19] to feel that they are incapable of making a significant impact,[20] and instead to focus on self-preservation to protect themselves from stress. To avoid this psychological backfire which will make learners vulnerable to misinformation, teachers who seek to address climate change in their classrooms must be very aware of what information is asserted by the narrative and what additional science stimuli can supplement this information during instruction and assessment.

Presenting a narrative that engages a range of human emotions, rather than only negative feelings, may help students recognize and face a crisis, so that students feel empowered to confront it. Indeed, the fear of potentially causing hopelessness or despair in students, due to the mainly negative feelings elicited by the climate crisis, is a major reason many teachers in the United States are uncomfortable about teaching climate change and, as a result, actively avoid doing so.[21] Providing teachers with an instructional option to frame climate change through a literary lens supports not only the well-being of students but also their ability to respond to this crisis.

Choosing an emotionally engaging narrative cannot be emphasized enough—this investment is critical for learning, not only helping learners better retain new information but allowing them to process information empathetically to promote prosocial action. Climate change requires communication, collaboration, creativity, and problem-solving between individuals and groups all over the world; therefore, solutions designed to address its many impacts first require empathy. Particularly in the medical field, empathy has been identified as a key component of providing excellent patient treatment, acting as a critical bridge between observation and action.[22] If we wish to create a similar bridge between awareness and climate action, empathy must be an essential aspect of knowledge sharing. Empathy is important for people to not only interact with civility, but also for learners to develop any twenty-first-century skills needed to address the impacts of climate change.

Social and emotional learning is becoming increasingly important to education leaders in the United States and abroad with a focus on trying to formalize SEL as part of curriculum, particularly after COVID-19 interrupted in-person instruction throughout the country and around the world.[23] In many cases, SEL curricula are disconnected from core discipline content and used more to support student achievement and promote a healthy school culture that extends to the wider community. However, in the case of teaching about climate change to support the development of informed, civically engaged students, pedagogy must foster empathy, one cornerstone of SEL, in concert with science education. Without including empathy as a key component of learning, how climate crisis affects certain communities in different ways will be poorly understood, making it difficult for learners to fully grasp the scope of the problem or how they might take action to address it.

Empathy is a key tenet of interpersonal awareness used to categorize SEL assessments,[24] suggesting how essential it is to the emotional state of a learner. This has perhaps never been more relevant as the world experiences COVID-19, and as America becomes acutely aware of the inequities the pandemic has exposed within its borders. Educating learners of all ages about climate change (which contributes to or exacerbates these inequities) requires the integration of an emotional-social component, especially considering that contributing factors are primarily caused by affluent communities from wealthy countries. These individuals, communities, and countries must take responsibility in recognizing how their actions disproportionately impact others not at fault for the climate crisis. However, admission requires an understanding that can be psychologically challenging, even threatening, especially to those who are unknowingly contributing to climate change, when empathy is not a component in reaching such recognition. Importantly for the classroom, this is not about assigning blame or creating guilt but rather building a bridge toward thoughtful perspective-taking that can help students understand certain realities inherent in the inequities of climate change and the myriad injustices that flow from it.

There are many psychological reasons for people to deny the existence of climate change, rationalize away its causes, or downplay its impacts, and there is no one method for unraveling deeply held beliefs that make someone comfortable with their perception of reality.[25] Empathy fosters a kind of introspection by trying as much as possible to understand, despite never fully experiencing, the perspectives and lived experiences of others. Educators wishing to explore this with their students might encourage the use of an empathy self-assessment scale as part of their evaluations of, and reflections on, learning. For example, engaging with a tool such as *The Empathy Continuum* created by IB educator Elizabeth Solomon invites students to identify their current stance on a particular local or global concern as well as what action they might take to deepen their understanding and move forward on the continuum.

Depending on the student's comfort level, interaction with the continuum might be done privately or could be shared as part of a group discussion. Either way, students could be encouraged to reflect on their position prior to, during, and after studying the concern so that they might consider how the learning experience impacts them; educators could collect this information anonymously to help inform future instructional strategies and content decisions. Student feedback about their position on the continuum at different stages of engagement with a topic could be a valuable tool for determining

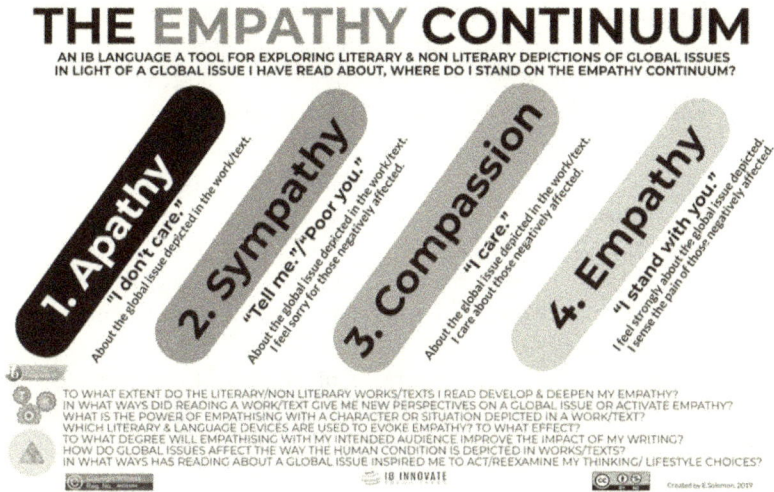

THE EMPATHY CONTINUUM

AN IB LANGUAGE A TOOL FOR EXPLORING LITERARY & NON LITERARY DEPICTIONS OF GLOBAL ISSUES
IN LIGHT OF A GLOBAL ISSUE I HAVE READ ABOUT, WHERE DO I STAND ON THE EMPATHY CONTINUUM?

1. Apathy "I don't care." About the global issue depicted in the work/text.

2. Sympathy "Tell me."/"poor you." About the global issue depicted in the work/text. I feel sorry for those negatively affected.

3. Compassion "I care." About the global issue depicted in the work/text. I care about those negatively affected.

4. Empathy "I stand with you." I feel strongly about the global issue depicted. I sense the pain of those negatively affected.

TO WHAT EXTENT DO THE LITERARY/NON LITERARY WORKS/TEXTS I READ DEVELOP & DEEPEN MY EMPATHY?
IN WHAT WAYS DID READING A WORK/TEXT GIVE ME NEW PERSPECTIVES ON A GLOBAL ISSUE OR ACTIVATE EMPATHY?
WHAT IS THE POWER OF EMPATHISING WITH A CHARACTER OR SITUATION DEPICTED IN A WORK/TEXT?
WHICH LITERARY & LANGUAGE DEVICES ARE USED TO EVOKE EMPATHY? TO WHAT EFFECT?
TO WHAT DEGREE WILL EMPATHISING WITH MY INTENDED AUDIENCE IMPROVE THE IMPACT OF MY WRITING?
HOW DO GLOBAL ISSUES AFFECT THE WAY THE HUMAN CONDITION IS DEPICTED IN WORKS/TEXTS?
IN WHAT WAYS HAS READING ABOUT A GLOBAL ISSUE INSPIRED ME TO ACT/REEXAMINE MY THINKING/ LIFESTYLE CHOICES?

IB INNOVATE   Created by E.Solomon, 2019

Figure 10.1. The Empathy Continuum.
Source: Elizabeth Solomon, 2019, IB-Innovate.com

not only prior knowledge and emotional readiness but also any potential biases and misunderstandings.

It is through a sense of empathy fostered by literature, coupled with rigorous science education, that climate change instruction can be truly meaningful. Psychological backfires that arise from exposure to narratives about climate change that are overwhelming, incomplete, or inaccurate, can be mitigated by teachers intentionally framing science content within a narrative that has specific qualities.

Drawing upon Nussbaum's[26] and others' analyses of narrative features that create empathy, teachers should select narratives that promote prosocial action and help counteract misinformation. These chosen narratives should

- include constructive solutions to the crisis that can motivate learners[27];
- offer some kind of "hook" that links information in the narrative to actual data or experiences;
- present in positive terms what people want the world to be like for future generations;
- tell a story that elicits a range of emotions from learners by demonstrating a range of character emotions;
- provide clear answers (as much as possible) for how a crisis arose to begin with, and specify if a crisis needs multifaceted and nuanced solutions; and
- offer an emotionally engaging storyline that addresses the complexities of the crisis rather than relying on it merely as a plot device.

Leveraging fictional worlds will generate student analysis of characters and settings that are grappling with environmental impacts. Engaging with actual and anticipated concerns presented by the stimuli set will spur investigative science opportunities that lead to creative solutions.

Measurements of learning can then consider students' understanding of and readiness to act upon a scenario established by the assessment task. Specifically, evaluation

- relies on the connection between storytelling and data analysis to illustrate complex patterns that exist within human/nonhuman behaviors and scientific phenomena;
- encourages the application of twenty-first-century learning skills through collaborative critical thinking, multidisciplinary approaches to problem-solving, and creativity in developing and communicating solutions;
- addresses standards from two or more disciplines to help demonstrate knowledge application across contexts;
- considers empathy as a motivator for engaging with stimuli from a perspective of self-reflection and prosocial action to encourage impact beyond the scope of the assessment; and
- provides a measurement of student progress in an individualized way that considers both standards-based performance expectations as well as twenty-first-century skills-based indicators of readiness for college/career/citizenry.

Considering these wide-ranging qualities, the process for selecting stimuli and creating assessments should be thoughtful. Additionally, the process must allow for flexibility based on an individual educator's objectives for any lesson, unit, course, and—significantly—district or program goals. In any case, the assessment should pair a narrative excerpt with a relevant data set that addresses a specific scientific phenomenon. Envisioning data-based solutions through the lens of a story allows students to interact with both the text and the data in a personal way.

The literary stimulus (fiction or nonfiction) could take a variety of storytelling forms such as narrative prose or poetry (including spoken word), a graphic novel, a web-based storybook, or a film or screenplay. Regardless of its form, the stimulus should focus on one or more science phenomena in the context of a conflict demonstrated by the narrative. It should be accompanied by questions that help illuminate how the conflict is portrayed and, if included, how its resolution is accomplished. In other words, an opportunity to study the author's craft. The nonliterary or data stimulus should be selected for its presentation of an aspect or aspects related to the science phenomena illustrated in the literary stimulus and might include a variety of data demonstrations, including simulations, graphics, survey or experiment results, or other formats as appropriate.

For example, one excerpt from Barbara Kingsolver's *Flight Behavior* explores the disruption of monarch butterflies' migratory patterns through the lens of Dellarobia Turnbow, the character who discovers them on her property. This passage might be accompanied by a map that tracks the traditional patterns monarchs follow or by a chart that illustrates their annual cycle of development and migration (including how each may be disrupted from year to year by environmental factors such as weather, the presence of pesticides, or the availability of food).

It is important to note that while we are advocating for assessments intended to address the complexities of climate change, this approach would be effective with any number of science phenomena. Additionally, the literary stimulus may be selected for its presentation of content relevant to another discipline such as history, the arts, mathematics, or economics. For this approach, data sets would be related to the specific discipline identified through the excerpt as has often been done on history or government exams that pair historical writings with primary source documents. The idea is that students may employ knowledge and skills from two (or more) disciplines to engage with the concept presented. First, students are expected to use the stimuli set to make sense of a phenomenon. Then, they should have options for presenting their understanding of it across these subject-specific standards. Where appropriate, presentations could be evaluated based on students' application of targeted twenty-first-century skills.

To foster meaningful engagement with the stimuli set, guiding questions that help students evaluate the data and explain the phenomenon from an interdisciplinary perspective should be offered. These may include variations on the following:

- How does this phenomenon affect humanity/ecosystems/Earth's systems?
- How has human activity contributed to this phenomenon, and how is this reflected in the literary stimulus?
- How is the presentation of the phenomenon in the literary stimulus different from or similar to the way it is presented by the nonliterary stimulus?
- To what extent does data from the nonliterary stimulus help inform the literary stimulus, and in what context does the literary stimulus explore information related to data in the nonliterary stimulus?

- How can this phenomenon be examined or explained to reflect both its scientific and societal importance?

Subject-specific scaffolding for each stimulus may also be appropriate to support understanding and encourage thoughtful reflection on each stimulus. These may be represented by guiding questions or by content engagement tasks:

- How is the scientific phenomenon represented by a conflict or conflicts in the literary stimulus?
- To what extent do the people or characters represented in the literary stimulus comprehend this phenomenon, and how is this demonstrated in their reactions to it?
- Analyze how the author's stylistic choices contribute to the portrayal of this phenomenon.
- Summarize the information presented in the nonliterary stimulus.
- Determine additional data sources that would be useful in further investigations of this phenomenon.
- Evaluate the information used to make predictions in the nonliterary stimulus.

Guiding questions and tasks may be evaluated against multidisciplinary standards or objectives or may simply be available to support students' progress toward a culminating performance task.

Whether parameters for the performance task may be defined by the nature of the assessment as either formative or summative, offering students a choice in how they demonstrate their learning could help encourage independence and confidence as well as authenticity in their approach. The following suggestions are meant to support educators' development of performance tasks that are suitable for their students (individually or in collaboration) and situations; they are not meant to exclude other valid options for demonstrating learning. Using the guiding stimuli set:

- Students are asked to explain or represent the phenomenon through the presentation of a student-selected medium (poetry, art, model, written or verbal analysis) or performance (a role-play, TED Talk, podcast) meant to impact their target audience.
- Students participate in a role-playing scenario in which they are asked to consider and identify which information may be needed for additional contextual understanding or multidisciplinary investigation (e.g., what evidence is absent for a scientific analysis, which stakeholders should be

involved in problem-solving, to what extent should external resources be employed toward proposed solutions?).

- Students engage in an actual investigative study that outlines what they observe about the scientific phenomenon, what information they consider valuable to gather for further analysis, how they will employ their collective content knowledge to evaluate the phenomenon, and how they will propose/develop potential approaches/solutions to problem-solving/capacity-building based on it.

As noted, more traditional questions may be included as part of the task to assist in illustrating comprehension aligned to grade-appropriate content standards of both the literary and nonliterary stimuli. These items may include short responses, highlighting and coding tasks, a comparison of how the stimuli present similar information in different ways, and a written or verbal interpretation of how the data in the nonliterary stimulus is represented in or by the literary stimulus (or vice versa).

An evaluation of the performance task should extend beyond traditional scoring approaches. Instead, these meta-assessments would support an individualized assessment of the student's performance across task items. They could include an evaluation of comprehension based on content standards, ability to engage collaborative skills toward collective problem-solving, effectiveness in communicating key ideas about the stimuli and scientific phenomenon, and ability to innovate clear solutions toward problem-solving and capacity-building. Importantly, the overall evaluation is focused on what the student can do with the information they have.

To provide this opportunity, part of the evaluation may include time for the student (or group of students if working collaboratively) to explain for evaluators how their chosen presentation media examines the scientific phenomenon and its relationship to the problem-solving they have engaged with; this may include discussing the limitations of their chosen media as well as the stimuli set. Depending on the twenty-first-century skills targeted in the performance task parameters, this may also be an opportunity for students to discuss how efficiently they feel they (and their peers if appropriate) managed their time, overcame specific obstacles, organized their decision-making, strategized solutions across disciplines, created their product, and communicated their ideas (including with one another if it was a group task) in their chosen media. Evaluators might consider subject-specific competencies including literary analysis and scientific accuracy, reflection of deeper understanding as it relates to the problem established by the stimuli set, and application of specific twenty-first-century skills (communication, creativity, critical thinking, collaboration, approach to interdisciplinary problem-solving) in the student's demonstration of learning. More traditional evaluations may also be included to assess standards-based performance in each discipline. However, scoring opportunities should include an individualized commentary on the student's ability to act on stimuli with independence and, if appropriate, in collaboration with others to show their appreciation for how one content area informs the other when envisioning potential investigations or solutions.

Cognia™,[28] a global education improvement organization, created an assessment model for this approach as part of an exploratory design process for developing innovative performance tasks. The exemplar pairs a short excerpt from Julie Bertagna's young adult novel *Exodus* with a data set that includes a sea-level rise and coastal flooding impact viewer provided by the National Oceanic and Atmospheric Administration (NOAA). These data sets collectively predict that sea-level rise will cause many parts of a selected coastal city to be at risk for being underwater by the end of this century, pairing well with the future dystopian setting established in the excerpt from *Exodus*; the first novel of a science fiction trilogy, the story begins on the island of Wing in the year 2100.

Selected for the concrete way Bertagna establishes conflict based on local and global sea-level rise through the perspective of a young female protagonist and other islanders, this literary lens illustrates not only the scientific phenomenon but different human perspectives and reactions toward it. Throughout her life Mara has been observing water levels overtake various landmarks on her small island and, in collaboration with an island elder named Tain, is now trying to persuade her community that they need to evacuate in search of higher ground. To her frustration, the other islanders perceive these anxious warnings as exaggerated and initially refuse to consider leaving their island homes. This is the opening conflict of the story that ultimately leads to the islanders' exodus from Wing. While the speculative plot that follows is highly engaging, students need only study this exposition to appreciate many of the complex real-world issues created by rising waters, particularly in threatened island nations and the many highly populated coastal cities around the world.

In addition to describing the physical characteristics of a changing landscape and raging sea, Bertagna explores the human psychological toll of trying to grasp and reconcile even the most concrete impacts of an altered environment. With sea levels literally encroaching on their homes and livelihoods, at one point even Mara seeks reassurance in the islanders' confidence that all will be well: "'People say it won't happen,' Mara bursts out. 'They say the ocean will settle again in the summer and we'll be safe.'" The archetypal voice of wisdom, Tain does not entertain this momentary doubt and instead confronts it head-on: "'But you can use your eyes, Mara, even if they can't!'

he cries. 'You're not a child anymore and I won't lie to you. We all have to open our eyes now . . . '"[29] As a voice of reason and truth, Tain's mentoring role is an interesting one to talk through with students to see how they associate him with people in their own lives, including parents, teachers, coaches, community, or religious leaders. Helping students determine who in their own lives may offer knowledgeable guidance in a time of conflict is another way educators can support students' exploration. Recognizing, of course, that this person may be the teacher, it is important to be prepared for conversations that extend beyond the text.

Though misguided in their refusal to acknowledge the evidence that exists all around them, the islanders' dismissal of Mara's and Tain's warnings represents common human responses to climate change: fear and denial. For the sake of young readers who may see themselves in the islanders, it is important to handle these characterizations with care; significantly, while the inhabitants of Wing are initially presented as antagonists who playfully mock and downplay the warnings Tain and Mara attempt, they are also revelers enjoying the momentous turn of a century who "just want to celebrate and forget." As one character pleads, "The children are all here. I don't want them frightened."[30] While a legitimate concern, the irony cannot be lost on readers who have seen a surge of student activism in the form of climate protests and youth climate organizations in recent years. Today's children are indeed frightened by the unfolding climate catastrophe, but they seek adults' participation and action, not their protection by denial. Through commentary on the islanders' perspective, the story's narrator establishes Wing in the broader context of this future world, thereby also critiquing another common response to climate change: "Earth may have abandoned others to its swallowing seas—people in far distant lands—but, they claim, that could never happen to *us*."[31] In likening the islanders' predicament to today's privileged, who presume the impacts of climate change will not affect them, Bertagna exposes a naivety that drives the inaction of those most culpable for the consequences millions around the world are currently experiencing. In an interesting case of life imitating art, Greta Thunberg has championed youth's collective call to action in demanding that adults take responsibility and recognize the realities that surround us. Like Mara, she has been ignored and even mocked for her youthful passion, yet she has persisted in the fight to spread awareness and urge change. The poet Eunice Andrada made a similar call to action in her performance of "Pacific Salt"[32] at the Paris Climate talks in 2015, which specifically addresses sea-level rise and its devastating impacts on Pacific Island Countries (PICs).

The nonliterary stimulus paired with an introductory excerpt from *Exodus* (although later excerpts would also work well) is an online NOAA tool that provides data on sea-level rise by location and timeline, with predictions of landscape impacts based on current rates of change in or near the locations specified. The timeline extension in the simulator is 2100, the same year of the novel's setting, offering a compelling connection and excellent opportunity for multidisciplinary analysis of impacts described in both the story and the simulator. Comparisons of consequences described in the fictional world with those looming in our own offer unique options for engaging students in deeper learning and innovative solution-finding. In the assessment design developed at Cognia™, tasks accompanying the stimulus set were targeted on the coastal area where students lived, making interaction with the data tool even more meaningfully relevant. For the excerpt selected, content-specific questions asked students to analyze evidence presented in the text in the form of details related to the islanders' plight. For example, in this description readers are shown the changing landscape Mara has witnessed throughout her young life:

> Mara's heart sinks even more as she looks across the field of whirling windmills and glinting solar panels to where there was once a long shoreline and road, a harbor, and the island's school. Just a few years ago, Mara and her friends went to the school. Then the sea claimed it. Even at midwinter you could still see its flat roof. Now it is completely lost.[33]

Analyzing the focus on imagery that demonstrates changes over time leads also to opportunities for reflection on the human markers that have been lost—road, harbor, school—and the power that is ascribed to nature in the simple statement "then the sea claimed it." An intriguing extension of this reflection might include excerpts from Alan Weisman's nonfiction *The World Without Us*, which imagines what might happen to the structural artifacts of humanity if people were suddenly to disappear from the earth and nature was permitted to take its course. The focus on details of time in the context of Mara's life also provides the opportunity for readers—likely of a similar age as Mara—to determine the potential impacts on their local shorelines, landscapes, and structures. For those not located near water, another specific coastal location could be selected for study. Using the NOAA tool, which provides a map simulation of the impact of increasing water levels, students investigate rates of sea-level rise for a particular area and determine potential effects on area infrastructure, including housing, businesses, transportation, or entertainment structures as well as natural landmarks that may be overcome. Depending on the area selected for the assessment, additional data sets could be included to support this activity such as population charts, detailed area maps, or possibly other data representations related to mitigation efforts already underway. Based on prior knowledge contexts and the specific objectives set for the assessment task, additional stimuli could be used to support students' understanding of the causes of sea-level rise, including animations, videos, and other visuals that help illustrate the concept of coastal erosion or allow students to investigate its complexities. Connecting projected impacts based on NOAA's (or another appropriate source's) simulator visualizations for a specific place with key descriptions presented in the story helps make otherwise abstract data more concrete. Numbers in a document may seem compelling but they cannot tell the story in quite the same way as the image of a beloved landmark disappearing in a sea-level rise simulation or as this narrative account of sea consuming land:

> To the north is a network of small, craggy islands. Once, they were all joined as a single landmass but over the last century the plains and much of the hills have been swallowed by sea and now only the peaks remain. Scattered across the slow-churning ocean, they look like bits of storm-tossed litter.[34]

Another very serious consequence of sea-level rise is addressed in the narrative's initial setting and develops as a central theme of the novel as Mara and the islanders themselves become refugees seeking higher ground. In the exposition, the narrator establishes credibility for Mara's concerns in further detailing the timeline of this speculative world while clearly indicating that the island can no longer accommodate its inhabitants:

> Over the last century many islanders have had to shift homes and farms and entire villages up out of reach of the rising ocean—some more than once. Wing, the largest and highest island, is now overcrowded with refugees from its northern neighbors, who have made makeshift homes in the ruins of its ancient stone cottages and farmyard outhouses.[35]

Obviously, accompanying data sets that provide information on current or predicted refugee circumstances of a particular area must be handled with care, but offered here is a clear opportunity to foster students' awareness and empathy for the very real plight of climate refugees, including implications on resources in the places to which they may migrate. Extension opportunities for further investigating this aspect of the climate crisis exist later in the novel itself when Mara and the islanders live for a time as climate refugees outside a walled city. For educators unsure of how to approach this reality in the classroom, nonfiction resources such as Jeff Goodell's *The Water Will Come: Rising Seas, Sinking Cities, and the Remaking of the Civilized World* offer accessible, relatable information that could inform pedagogical approaches.

In planning assessment formats that are based on pairing *Exodus* excerpts and appropriate science stimuli, a teacher could develop a progression of formative evaluations in a scaffolded manner, starting with assessments that target reading standards, followed by multidisciplinary assessment of Earth science with reading, engineering and technology standards. Assessments could begin by focusing on the evidence of sea-level rise in the narrative before having students connect the details of sea-level rise to actual data, using stimuli such as simulators, maps, or graphs that illustrate predictions for a particular area. As just a few examples, teachers could develop assessment tasks where students evaluate potential solutions to flooding while making connections between what occurs in *Exodus* with predicted events from data sets, or where students connect the behavior of characters in *Exodus* to human perspectives related to long-term coastal erosion. For example, there are many video narratives documenting the devastating impact of sea-level rise on the people of Kiribati. The progression could then end with a culminating performance task that integrates assessment of Earth science, engineering and technology standards with reading, writing, and speaking standards while incorporating twenty-first-century skills as part of the demonstration of learning. This performance task prompt developed by Cognia™ is one example:

> Based on all the information presented to you and your answers to the questions about *Exodus*, prepare and give a presentation for your community that urges them to act now to protect your community from the negative future effects of sea-level rise.[36]

This approach is especially compelling since the main protagonist of *Exodus*, Mara, is herself a young adult trying to convince her community to address an ongoing and worsening crisis. Students would essentially relate to the narrative as if they were the main character, making connections between the fictional world and their own reality. At the same time, teachers could specify criteria for this presentation so that students integrate understanding of science concepts, reading comprehension, and solutions development into one cross-curricular task, making this a comprehensive assessment for teachers to support student engagement, communication, creativity, and critical thinking in understanding and addressing real crises.

For example, in assessing a student's culminating performance task centered on *Exodus*, teachers could design rubrics to score how a student problem-solves and makes connections between science and literature in addressing sea-level rise. This sample rubric developed by Cognia™ shows how multidisciplinary critical thinking and problem-solving skills could be generally evaluated and customized for any science-literature pairing:

Table 10.1. Evaluation Criteria: Multidisciplinary Critical Thinking and Problem-Solving. Source: Cognia™

Level	Performance Description	Task-Specific Objectives
4	The performance demonstrates **a sophisticated** multidisciplinary approach to the task.	-The presentation addresses the audience by making **thorough connections** between the scientific data and literary context. -The student **synthesizes** content information to offer unique, **comprehensive** solutions that are stimuli-inspired but may extend beyond those presented in the assessment.
3	The performance demonstrates an **effective** multidisciplinary approach to the task.	The presentation addresses the audience by making **clear connections** between the scientific data and literary. -The student **analyzes** content information to offer **thoughtful**, stimuli-inspired solutions.
2	The performance demonstrates an **adequate** multidisciplinary approach to the task.	The presentation addresses the audience by making **sufficient** connections between the scientific data and literary context. The student **considers** the content information to offer **standard**, stimuli-supported solutions.
1	The performance demonstrates a **limited** multidisciplinary approach to the task.	The presentation addresses the audience by making **minimal** connections between the scientific data and literary context. -The student **summarizes** the content information to offer **basic** and potentially **incomplete** stimuli-related solutions.
0	The student **does not** demonstrate a multidisciplinary approach to the task.	No response or response is inappropriate for the task.

To further assess twenty-first-century skills used in a culminating performance task, rubrics could also evaluate how effectively a student develops and presents an argument for addressing sea-level rise. This sample rubric developed by Cognia™ shows how such planning and communication criteria could be evaluated:

Table 10.2. Evaluation Criteria: Planning and Communication. Source: Cognia™

Level	Performance Description	Task-Specific Objectives
4	The performance demonstrates an **original** approach to the task.	-The presentation addresses the audience with confidence, showing independence of thought and **excellent** clarity of expression. -The student makes **persuasive** stylistic and content choices for conveying solutions and engaging the audience.

Level	Performance Description	Task-Specific Objectives
		-The student organizes information in a compelling manner, demonstrating appreciation for how the information **connects** across a variety of contexts (i.e., structural, behavioral, historical, financial, literary, scientific).
3	The performance demonstrates a **competent** approach to the task.	-The presentation addresses the audience with confidence, showing independence of thought and **mostly clear** expression.
		-The student makes **effective** stylistic and content choices for conveying solutions and persuading the audience.
		-The student organizes information in an effective manner, demonstrating awareness of how the information **relates** to different contexts.
2	The performance demonstrates a **fair** approach to the task.	-The presentation addresses the audience with some confidence, showing some independence of thought and **basically clear** expression.
		—The student makes **good** stylistic and content choices for conveying ideas and informing the audience.
		-The student organizes information in an appropriate manner, demonstrating some awareness of how the information **applies** to the literary and scientific contexts.
1	The performance demonstrates a **weak** approach to the task.	-The presentation addresses the audience with limited confidence, showing little independence of thought and **weak** expression.
		—The student makes **basic** stylistic and content choices for conveying ideas to the audience.
		-The student organizes a few pieces of information that **relate** to the literary or scientific context.
0	The performance **does not** demonstrate a creative or persuasive approach to the task.	No response or response is inappropriate for the task.

By having students develop this presentation with the goal of convincing people in their community to take prosocial action on sea-level rise, students would be developing their own conceptual approach to their presentation while simultaneously processing the perspectives of other people. To craft a convincing argument, students would need to connect the points of view of the characters in *Exodus* to those of real people, thus providing an opportunity to incorporate community interviews as part of the task. During this process students will demonstrate not only their understanding of the science behind the phenomenon but also how to communicate the science in a way that helps other people understand it accurately enough to feel a sense of urgency to act.

Students who successfully perform this culminating task may be in a better position to become their own teachers and self-assessors because they are aligning their understanding of the world with the different perspectives of the people they teach, thereby reflecting on their own metacognition. Learning how to teach and assess themselves is a critical skill that would allow students to continuously check their own understanding and seek additional knowledge.

Although climate change is a complex and multifaceted topic, with many interconnected scientific phenomena, it is imperative that teachers recognize the necessity of preparing today's learners and take appropriate steps toward providing them with the knowledge and critical skills they will need to be innovative, proactive, and resilient in the face of a crisis that demands it.

To help teachers take the first step in this pedagogical journey, Table 10.3 compiles a list of some environmental and climate change-related topics that are paired with literature appropriate for different ability ranges. These connections and topics are not exhaustive but provide starting points for teachers to explore climate change instruction with this approach.[37]

Educating about climate change only in science classes or elective courses is no longer an option. Nor are assessments that measure isolated content-based skills rather than essential soft skills that can indeed be taught, practiced, and mastered. What it means to be college and career ready needs to be drastically reimagined. Students at every age level deserve an education that adequately prepares them for the choices they will be making (indeed must make) in a climate-challenged world. They deserve to understand not only the severity and complexity of the problem but also the multidisciplinary, interrelated nature of its potential solutions.

A learning experience designed with this goal in mind provides valuable insights into students' level of readiness to act in several key areas. No longer singularly content or standards-based, such an evaluation offers educators information about how well students are prepared to respond to actual challenges by determining how effectively they can cooperate, problem-solve, create, and communicate. Learners of all ages must achieve climate literacy if we hope to equip them for confronting the challenges this century brings. Providing authentic learning experiences that are more personally relevant, meaningful, and engaging for all learners is, therefore, a critical step. Let us create culturally responsive assessments that value the skills students need in a climate-changed world and that prioritize the agency students deserve to confront the myriad injustices exacerbated by this crisis.

Table 10.3. Environmental and Climate Change-Related Topics Paired with Fiction. Source: Cognia™

Environmental Topic	Literature Connections	Key Subject(s)	Target Grade Level
Pollution			
Pollution in general	*Eco-Wolf and the Three Pigs*	Environmental twist on a classic	Elementary
Plastic pollution	*Alba and the Ocean Clean Up*	Ocean, plastic pollution, coral reefs	Elementary
Waste disposal	*What a Waste! Where Does Garbage Go?*	Garbage, human waste, food waste, waste disposal systems	Elementary
Oil spill pollution	*Flight or Fight*	Vancouver, Canada; seabirds	Middle or high school
	How Beautiful We Were	Fictional African setting; environmental degradation related to oil	
Ocean pollution	*They Came from Below*	Cape Code, global pollution	Middle or high school
	Washashore	Fragile shorelines; ospreys	
Recycling	*Saving the Planet and Stuff*	Inner workings of environmental magazine activism	High school
DDT contamination and effect on organisms	*Mosquito Point*	Boat building, spraying of DDT around a bay	High school
	"A Fable for Tomorrow" (in *Silent Spring*)	Impact of DDT on bird populations	High school
Conservation			
Effects of logging, mining, land development, and fracking	*The Ghost Runner* (Lithia Trilogy, book 2)	Logging and gold mining	High school
	A Bird, a Girl, and a Rescue (The Rwendigo Tales, book 2)	Africa, human trafficking, illegal logging	Middle school
	The True Blue Scouts of Sugar Man Swamp	Bayou, feral pigs, land development	Elementary and middle school
	The Earth Is My Mother	Native American protagonist, southwestern United States	Middle and high school
	The End of the Wild	Fracking, land development	Elementary
Protecting habitat of burrowing owls	*Hoot*	Southern Florida, children try to save wildlife	Middle school

Shark conservation	If Sharks Disappeared	Importance of sharks to ecosystems, impacts of overfishing, how to help sharks	Elementary
Protecting endangered species	The Adventures of the Sizzling Six Series	Teenagers on an ecoadventure to save an endangered species	Middle and high school
	The Case of the Missing Cutthroats	Cutthroat trout, eco-mystery	Middle and high school
	The Last Panther	Climate refugees, walled cities, habitat loss, species extinction	Middle and high school
Effect of hunting on ecosystems	Lostman's River	1900s, Florida Everglades, thirteen-year-old protects environment	Middle and high school
South American rainforests	Rogue Harvest	Dystopian future Earth, eco-mystery	Middle and high school
Deforestation	A Forest, a Flood, and an Unlikely Star	Africa, AIDS crisis, endangered gorillas	High school
Africa's landscape	A Chameleon, a Boy, and a Quest	Africa, talking chameleon, orphan protagonist	Middle and high school
Restoring natural environments	Green Boy	Mute protagonist, Bahamas, fantasy world	Middle and high school
Ecological balance	The Missing "Gator of Gumbo Limbo"	Homeless protagonist, southern Florida, alligators	Middle and high school

Environmental Events

Drought	Parched *Dry* The Water Knife *Gold Fame Citrus*	Dystopian future, resource availability, water crises	High school
Ecological disaster	Timelock (third book in trilogy after Rogue Harvest and *Blood Sun*) The Fifth Season (first book of The Broken Earth series) *Orleans*	Dystopian, past/future Earth Dying land, water scarcity, competition for resources Gulf coast, superstorms, disease outbreak	Middle and high school High school

Environmental Topic	Literature Connections	Key Subject(s)	Target Grade Level
Natural forest fires and ecosystems	*Fire Birds: Valuing Natural Wildfires and Burned Forests*	Bird species that rely on burned forests, fire policies, counters negative view of forest fires	Elementary and middle school
Volcanic eruptions	Trilogy: -*Ashfall* -*Ashen Winter* -*Sunrise*	Yellowstone, survival in transformed landscape	Middle and high school
Melting of ice caps	*The Big Melt*	Effect of climate catastrophes on a personal level	Middle and high school
Global warming	*Solstice*	Fantasy, addressing crises through agency	Middle and high school
Intense storms due to climate change	*The Carbon Diaries 2015*	Dystopian future Earth, UK, carbon rationing, eco-thriller	Middle and high school
	Empty	Dystopian future Earth, oil supply runs out, students lead environmental reforms in society	High school
	Vigil Harbor	Superstorms, climate instability	High school
Citizen science	*The Fire Bug Connection*	European fire bugs, scientific reasoning, ecomystery	Middle and high school

NOTES

1. National Research Council, *Assessing 21st-Century Skills: Summary of a Workshop.* J. A. Koenig, Rapporteur. Committee on the Assessment of 21st-Century Skills. Board on Testing and Assessment, Division of Behavioral and Social Sciences and Education. Washington, DC: The National Academies Press, 2011. Introduction, https://www.nap.edu/catalog/13215/assessing-21st-century-skills-summary-of-a-workshop.

2. National Research Council, 2.

3. National Research Council, 9.

4. National Research Council, 11.

5. Tony Wagner, *Creating Innovators: The Making of Young People Who Will Change the World.* (New York: Scribner, 2012), 142.

6. Wagner, 200.

7. Jenny Nagaoka, Camille A. Farrington, Stacy B. Ehrlich, and Ryan D. Heath, *Foundations for Young Adult Success: A Developmental Framework. Concept Paper for Research and Practice* (University of Chicago Consortium on Chicago School Research, 2015), https://files.eric.ed.gov/fulltext/ED559970.pdf.

8. Romesh Gunesekera, *Reef.* (Penguin Books, 2014), 60.

9. Page Keeley, *Science Formative Assessment* (Corwin Press, 2008).

10. Keeley.

11. Jeremy Taylor, Katie Buckley, Laura S. Hamilton, Brian M. Stecher, Lindsay Read, and Jonathan Schweig, *Choosing and Using SEL Competency Assessments: What Schools and Districts Need to Know* (Rand Corporation, November 2018), practitioner-guidance.pdf (casel.org).

12. J. W. Pellegrino, M. Wilson, J. Koenig, and A. Beatty, eds., *Developing Assessments for the Next Generation Science Standards.* (Washington, DC: National Academies Press, 2014).

13. Lisa Zunshine, *Why We Read Fiction: Theory of Mind and the Novel* (Columbus: The Ohio State University Press, 2006), 6.

14. Keith Oatley, "Theory of Mind and Theories of Minds in Literature," in *Theory of Mind and Literature*, eds. Paula Leverage, Howard Mancing, Richard Schweickhert, and Jennifer Marston William (Purdue University Press, 2011), 14.

15. National Research Council, 42.

16. Note that adaptations should be made for students with autism.

17. R. T. Hartle, S. Baviskar, and R. Smith, "A Field Guide to Constructivism in the College Science Classroom: Four Essential Criteria and a Guide to Their Usage," *Bioscene* 38 (2012), https://files.eric.ed.gov/fulltext/EJ1002158.pdf.

18. Marilee Sprenger, *Learning and Memory: The Brain in Action.* Association for Supervision and Curriculum Development, 1999.

19. Paul Slovic, "If I Look at the Mass I Will Never Act: Psychic Numbing and Genocide," in *Emotions and Risky Technologies,* ed. S. Roeser (The International Library of Ethics, Law and Technology, 2010): vol. 5, 37–59, https://doi.org/10.1007/978-90-481-8647-1_3

20. Daniele Västfjäll, Paul Slovic, and Marcus Mayorga, "Pseudoinefficacy: Negative Feelings from Children Who Cannot Be Helped Reduce Warm Glow for Children Who Can Be Helped," *Frontiers in Psychology* 6 (2015): 61. https://www.frontiersin.org/articles/10.3389/fpsyg.2015.00616/full.

21. Anya Kamenetz, "Most Teachers Don't Teach Climate Change; 4 in 5 Parents Wish They Did," NPR, April 22, 2019, https://www.npr.org/2019/04/22/714262267/most-teachers-dont-teach-climate-change-4-in-5-parents-wish-they-did.

22. H. Riess, "The Science of Empathy." *Journal of Patient Experience* (June 2017): 74–77. doi:10.1177/2374373517699267

23. Matthew N. Atwell and John N. Bridgeland, "Ready to Lead: A 2019 Update of Principals' Perspectives on How Social and Emotional Learning Can Prepare Children and Transform Schools." A Report for CASEL. 2019. https://eric.ed.gov/?id=ED602977.

24. Jeremy J. Taylor, Katie Buckley, Laura S. Hamilton, Brian M. Stecher, Lindsay Read, and Jonathan Schweig.

25. See, for example, George Marshall, *Don't Even Think About It: Why Our Brains Are Wired to Ignore Climate Change* (Bloomsbury, 2014)

26. Martha Nussbaum, "The Paradox of Narrative Empathy and the Form of the Novel, or What George Eliot Knew," *Studies in the Novel* 48, no. 1 (2016): 19–42. See introduction for a summary of features.

27. Saffron J. O'Neill, Maxwell Boykoff, Simon Niemeyer, and Sophie A. Day, "On the Use of Imagery for Climate Change Engagement." *Global Environmental Change* 23, no. 2 (2013): 413–21. ISSN 0959–3780, https://doi.org/10.1016/j.gloenvcha.2012.11.006.

28. A nonprofit located in Georgia, USA.

29. Julie Bertagna, *Exodus* (London: Picador, 2002), 13.

30. Bertagna, 9.

31. Bertagna, 10.

32. Eunice Andrada, "Pacific Salt." https://www.youtube.com/watch?v=T4C2g-PHE4Q.

33. Bertagna, 14.

34. Bertagna, 14.

35. Bertagna, 14.

36. This prompt is from an innovative science performance task design developed by Cognia™.

37. See the Climate Lit initiative at ClimateLit.Org for more examples.

BIBLIOGRAPHY

Andrada, Eunice. "Pacific Salt." https://www.youtube.com/watch?v=T4C2g-PHE4Q.

Atwell, Matthew N., and John N. Bridgeland. Ready to Lead: A 2019 Update of Principals' Perspectives on How Social and Emotional Learning Can Prepare Children and Transform Schools. A Report for CASEL. 2019. https://eric.ed.gov/?id=ED602977.

Bertagna, Julie. *Exodus*. London: Picador, 2002.

Gunesekera, Romesh. *Reef*. Penguin Books, 2014.

Hartle, R. Todd, Sandhya Baviskar, and Rosemary Smith. "A Field Guide to Constructivism in the College Science Classroom: Four Essential Criteria and a Guide to Their Usage," *Bioscene* 38 (2012): 31–34. https://files.eric.ed.gov/fulltext/EJ1002158.pdf.

Kamenetz, Anya. "Most Teachers Don't Teach Climate Change; 4 in 5 Parents Wish They Did." NPR. April 22, 2019. https://www.npr.org/2019/04/22/714262267/most-teachers-dont-teach-climate-change-4-in-5-parents-wish-they-did.

Keeley, Page. *Science Formative Assessment*. Corwin Press, 2008.

Lee, Ohkee. Making Everyday Phenomena Phenomenal: Using phenomena to promote equity in science instruction. *Science and Children* 58 (September/October 2020:1. https://www.nsta.org/science-and-children/science-and-children-septemberoctober-2020/making-everyday-phenomena

Nagaoka, J., C. A. Farrington, S. B. Ehrlich, and R. D. Heath. (2015). *Foundations for Young Adult Success: A Developmental Framework*. Chicago: University of Chicago Consortium on Chicago School Research.

National Research Council. *Assessing 21st-Century Skills: Summary of a Workshop*. J.A. Koenig, Rapporteur. Committee on the Assessment of 21st-Century Skills. Board on Testing and Assessment, Division of Behavioral and Social Sciences and Education. Washington, DC: The National Academies Press, 2011. https://www.nap.edu/catalog/13215/assessing-21st-century-skills-summary-of-a-workshop.

Nussbaum, Martha. "The Paradox of Narrative Empathy and the Form of the Novel, or What George Eliot Knew." *Studies in the Novel* 48, no. 1 (2016): 19–42.

Oatley, Keith. "Theory of Mind and Theories of Minds in Literature," in *Theory of Mind and Literature*. Edited by Paula Leverage, Howard Mancing, Richard Schweickhert, and Jennifer Marston William, 13–26. Purdue University Press, 2011.

O'Neill, Saffron J., Maxwell Boykoff, Simon Niemeyer, and Sophie A. Day. "On the Use of Imagery for Climate Change Engagement." *Global Environmental Change* 23, no. 2 (2013): 413–21. ISSN 0959–3780, https://doi.org/10.1016/j.gloenvcha.2012.11.006.

Pellegrino, J. W., M. Wilson, J. Koenig, and A. Beatty (Eds.) *Developing Assessments for the Next Generation Science Standards*. Washington, DC: National Academies Press, 2014.

Riess, H. "The Science of Empathy." *Journal of Patient Experience* (June 2017): 74–77. doi:10.1177/2374373517699267.

Slovic, Paul. "If I Look at the Mass I Will Never Act: Psychic Numbing and Genocide," in *Emotions and Risky Technologies*, edited by S. Roeser, 37–59. The International Library of Ethics, Law and Technology, 2010: 5. https://doi.org/10.1007/978-90-481-8647-1_3

Sprenger, Marilee. *Learning and Memory: The Brain in Action*. Association for Supervision and Curriculum Development, 1999.

Taylor, Jeremy J., Katie Buckley, Laura S. Hamilton, Brian M. Stecher, Linsay Read, and Jonathan Schweig. *Choosing and Using SEL Competency Assessments: What*

Schools and Districts Need to Know. Rand Corporation, November 2018. practitioner-guidance.pdf (casel.org)

Västfjäll, Daniel, Paul Slovic, and Marcus Mayorga. "Pseudoinefficacy: Negative Feelings from Children Who Cannot Be Helped Reduce Warm Glow for Children Who Can Be Helped." *Frontiers in Psychology* 6 (2015): 61. https://www.frontiersin.org/articles/10.3389/fpsyg.2015.00616/full

Wagner, Tony. *Creating Innovators: The Making of Young People Who Will Change the World*. New York: Scribner, 2012.

Young, Rebecca. *Confronting Climate Crises through Education: Reading Our Way Forward*. Lanham: Lexington Books, 2018.

Zunshine, Lisa. *Why We Read Fiction: Theory of Mind and the Novel*. Columbus: The Ohio State University Press, 2006.

Afterword

Vandana Singh

When I first started teaching the basic science of climate change in my general physics classes, I found that contrary to my intent, I was unable to inspire students, let alone empower them for action. Later, I discovered that my experience in the microcosm of my classroom was reflected on a global scale—education has not risen to the challenge of the climate crisis, and one reason for this is the lack of radical visions for climate pedagogy.[1] My failure led me to an ongoing journey of exploration, developing a transdisciplinary pedagogy of climate change,[2] in which storytelling is central. As a writer of speculative fiction, I am especially aware of the power of a good story in the context of possible futures. Stories of various kinds—fictional, real-life, including dramatizations of scientific climate processes—have helped my students transcend multiple boundaries—between disciplines, between the personal and political, between worlds of possibility.

As a species that makes sense of the world through narrative, we live by stories, explicit and implicit. Advertisers and businesses have known this for a long time. Multiple disciplines have also recognized the power of narrative—for example, narrative medicine,[3] and the recent epistemological expansion in climate science that is inclusive of various forms of narrative (such as physical climate storylining[4]). It is more than high time that educators from all disciplines paid attention to storytelling in the classroom as a valuable tool for climate education.

In their groundbreaking book, *Storylistening: Narrative Literacy and Public Reasoning*, Sarah Dillon and Claire Craig make a careful case for the need for narrative literacy. They point out that the greatest power of well-chosen stories is potentially in the presentation of multiple points of view. As Chimimanda Adichie so forcefully points out in her TED Talk "The Danger of a Single Story,"[5] we need multiple narratives and narratives

231

that present multiple perspectives. Dillon and Craig also point to other key functions of stories—they help create identity; they are ontological tools for making sense of the world. The authors discuss *dominant narratives*, which are often invisible—our default assumptions about the good life, the nature of work, our economic and social arrangements, human exceptionalism, hierarchies of race, gender and caste. These narratives become templates for popular television shows, for example, that elevate, reinforce and normalize such default assumptions. Unexamined cultural norms are shored up through such dominant narratives. As a speculative fiction writer from India living in the United States, I am aware that dominant narratives push aside other perspectives, paradigms, and stories through new forms of colonialism. Our globalized socioeconomic system propagates these stories around the world: that the good life is synonymous with ever-increasing material and energy consumption, that there is an inherent separation between humans and nature, that technology or the market will fix all our problems, that the super-rich are to be admired, that humans are inherently destructive, that to solve the climate crisis a sacrifice is necessary, and the like. Such dominant narratives help shore up a socioeconomic system that is inherently unequal, and which, it can be argued, is the root cause of the climate crisis. Those who are most invested in such a system—the rich, the powerful, and the privileged—are least likely to want to change it. Those who have less power—people of color, women, Indigenous people, the young—are not only likely to have experienced the seamy underbelly of modern industrial civilization, but are more likely to want to change these systems of oppression and destruction. They are the ones whose stories, experiences, and epistemologies are not heard, not valued, and most needed. Carefully chosen stories, fictional and real-world, can help counter dominant narratives by presenting multiple alternative perspectives, including those of marginalized peoples whose stories give rise to new (to the rest of us) insights, perspective-shifting concepts, and novel ways to engage meaningfully with our crises—stories in which the nonhuman speaks, stories in which the Other becomes the center, stories in which we interrogate power, then become ontological tools for constructing alternative worlds. Speculative fiction, in particular, has the potential for breaking us out of the imagination trap by making default, taken-for-granted aspects of the world visible and contestable, and by immersing us in alternative realities. When such fiction is informed by real-world experiments at the grassroots, such as those recorded in the Vikalp Sangam project in India,[6] then it can powerfully reorient the imagination toward viable, just, and ecologically sound futures.

It is my sincere hope that the essays in this volume will add to, inspire, and provoke more attention to narrative and storytelling as part of a meaningful pedagogy of climate change.

NOTES

1. Christina Kwauk, "Roadmaps to Quality Education in a Time of Climate Change,"
[Brief] (Brookings Institute, 2020), https://www.brookings.edu/wp-content/uploads/2020/02/Roadblocks-to-quality-education-in-a-time-of-climate-change-FINAL.pdf.

2. Vandana Singh, "Toward a Transdisciplinary, Justice Centered Pedagogy of Climate Change," in *Curriculum and Learning for Climate Action: Toward an SDG 4.7 Roadmap for Systems Change*, ed. C. Kwauk and Radhika Iyengar, (UNESCO-IBE, 2021), https://brill.com/view/title/60973.

3. See, for example, https://www.vox.com/the-highlight/2020/2/27/21152916/rita-charon-narrative-medicine-health-care

4. Theodore G. Shepherd and A. Elisabeth, "Meaningful Climate Science," *Climatic Change* 169, no. 17 (2021), https://doi.org/10.1007/s10584-021-03246-2.

5. Chimamanda Ngozi Adichie, "The Danger of a Single Story," filmed 2009. TEDGlobal, 18:33, https://www.ted.com/talks/chimamanda_ngozi_adichie_the_danger_of_a_single_story?language=en.

6. See https://vikalpsangam.org/.

BIBLIOGRAPHY

Adichie, Chimamanda Ngozi. "The Danger of a Single Story." Filmed 2009. TEDGlobal, 18:33, https://www.ted.com/talks/chimamanda_ngozi_adichie_the_danger_of_a_single_story?language=en.

Dillon, S., and C. Craig. *Storylistening: Narrative Evidence and Public Reasoning*. Routledge, 2022.

Kwauk, C. "Roadmaps to Quality Education in a Time of Climate Change" [Brief]. Brookings Institute, 2020. https://www.brookings.edu/wp-content/uploads/2020/02/Roadblocks-to-quality-education-in-a-time-of-climate-change-FINAL.pdf.

Shepherd, T.G., and E. A. Lloyd. "Meaningful Climate Science." *Climatic Change* 169, no. 17 (2021). https://doi.org/10.1007/s10584-021-03246-2

Singh, V. "Toward a Transdisciplinary, Justice Centered Pedagogy of Climate Change." In C. Kwauk and Radhika Iyengar (eds.), *Curriculum and Learning for Climate Action: Toward an SDG 4.7 Roadmap for Systems Change*. (2021) UNESCO-IBE. https://brill.com/view/title/60973.

Vikalp Sangam (Alternatives India). https://vikalpsangam.org/.

Index

Abdalati, Waleed, 50
active debris removal (ADR), 53–54
activism (environmental, climate-related): political, 115–17, 181; student/youth, 9, 12, 14, 19, 22–23, 113–14, 119–21, 123–27, 133, 217, 224. *See also* agency
Adichie, Chimimanda, 231
advocacy: animal, 153; citizen, 50, 54, 64; climate change, 31, 97; environmental, 29, 76, 188–89, 179, 191–92; student, 66
Affective Ecologies: Empathy, Emotion, and Environmental Narrative, 2
agency: plant, 142, 152; student, 10, 12–13, 19, 38, 86, 115–17, 207, 223. *See also* activism
Andrada, Eunice, 217
anthropocentric ethos/culture, 77–78, 85, 87
anti-bias, 96
Ariely, Dan, 33
Attenborough, David, Sir, 130

Baker, David, 58
bees: African stingless, 35; interdependence of and critical role of, 105–7; threats to, 107
Behind the Curve, 38–44

Bennett, Jeffrey, 56–57, 67
biocentric, 78
biodiversity, 31, 34, 37, 59, 166, 188, 191
biomimicry, 65
biosphere, 50
Breath, 122
Bugs, 31–38

Canadian Space Agency, 67
Capitalism: A Ghost Story, 183, 186, 190
carbon footprint, 101, 118, 133
Carson, Rachel, 75–76
Chasing Coral, 204
Cherry, Kendra, 32–33, 38
"A Choking Sky," 122
The City of Ember, 96, 97–99, 101–3, 105–6
city zoning, 97
climate change: advocacy related to, 95–96, 114, 116, 201; impacts and threats of, 96, 103–4, 107, 128, 138, 178, 223; injustice and inequities of, 83, 158, 161, 189, 210, 223; politics of, 178–79; raising student awareness of, 86–90, 109, 117–18, 122–24, 126, 128–29, 153–54, 178, 223. *See also* climate education

climate-conscious (decisions, decision-making), 97, 104, 109
climate denial, 1, 20, 39–42, 44, 89, 210, 217
climate education: literacy and pedagogy, 128–29, 138–51, 178–80, 192–93, 201, 203–7, 208–10, 223 231–32
climate fiction, 211, 217, 224–26
climate justice, 9–10, 12–13, 15–16, 21, 23
climate refugee(s), 219, 225. *See also* displacement, climate-related
climate science, 1, 40, 230–31; scientists, 1–2
coding, 100, 102, 106–7, 215
cognitive bias, 32–32, 39, 42–43
cognitive dissonance, 35, 41
colonialism, 10, 232; colonial trade, 184, 186, 190, 197n59
congested parking garage effect, 53
conservation, 224, 190
conspiracy theories, 39–40, 42–44
consumerism (anti), 155–56
COVID-19, 32, 81, 86, 209–10
Craig, Claire, 231–32
Creating Innovators: The Making of Young People Who Will Change the World, 203
Crossan, Sarah, 122

"The Danger of a Single Story," 231–32
The Day After Tomorrow, 2–3
DeLillo, Don, 76, 81, 83, 85
denialism, 40–42, 89
Dennis, Carl, 127
Dillon, Sarah, 231–32
di Pippo, Simonetta, 49, 67
displacement, climate-related, 181–82, 187–88, 190. *See also* climate refugee(s)
documentary (film), 29–31, 36, 39–44, 80, 107, 125–26, 129–33
Don't Look Up, 1–3
Dunning-Krueger effect, 42

DuPrau, Jeanne, 96
dystopia: corporate, 190; dystopian setting, 208, 216, 225

ecocritical: analysis in video games, 81, 83, 90–91; approach and models, 75–76, 79–81; awareness, 179–80; studies in film, 80; studies in literature, 76, 79, 83, 122, 177–79, 192
ecocriticism: definition and history of, 78, 122; field or discipline of, 75–76, 78–81, 89–91, 178, 192; postcolonial, 177, 188–89; value of in teaching, 13, 119–20, 128, 138
The Ecocriticism Reader, 78
ecofriendly, 90. *See also* climate education, literacy and pedagogy
ecology, science of, 79
economy, 35, 58–59, 84, 96, 105, 107
ecopedagogy. *See* climate education, literacy and pedagogy
ecosystem(s): disruptions or threats to, 166, 177, 188, 191; educating about, 13, 99, 119, 204–5, 213; importance and conservation of, 32, 35, 145; stories about, 205, 224–26
Emmerich, Roland, 3
empathy: toward characters and in education, 2, 40–43, 55–56, 125, 204, 206–7, 209–12, 219; *The Empathy Continuum*, 210
environment(s): natural, 78–86, 89, 106, 113, 116–18, 121–23, 126, 128, 138–39, 143, 145, 156, 161, 165, 213, 225; school, 10, 202, 206; space, 49–50, 52, 56–58, 62–66
environmental advocacy, 4, 29, 188, 191–92
environmental awareness, 75–76, 131, 153
environmental education, 49, 55. *See also* climate education, literacy and pedagogy

environmental justice, 13, 88, 107, 189.
  See also climate change, injustice
  and inequities of
"Epithalamia," 127
erosion (coastal) 96, 98, 102–4,
  205, 219–20
extinction, 84, 89, 180, 191, 225
European Space Agency, 67
*Everything Change: An Anthology of
  Climate Fiction*, 122, 128

falsifiable hypothesis, 42–43
The Fate of the World, 81
flat-Earther, 40–44
Flood of Fire, 185–86
Food and Agriculture Organization
  of the United Nations (FAO), 31,
  34, 35, 36
Freire, Paulo, 16

gaming, 89–90
Geckos, 64
geckos, 54, 64–65
geostationary orbit (GEO), 51
Ghosh, Amitav, 139, 177–86, 188–93
global food production, 31–32, 106–7
global positioning system (GPS), 50
global warming, 53–54, 57, 67,
  89–90, 95, 137–38, 142, 146, 152,
  191, 203, 226
Glotfelty, Cheryll, 78
Goddard, Robert H., 55, 67
Goodell, Jeff, 219
Gottschall, Jonathan, 3
*The Great Derangement: Climate
  Change and the Unthinkable*, 178
greenhouse effect, 51, 53
"The Greenhouse Effect," 127

Hadfield, Chris, 62
Hansen, James, 95
The Happening, 137–48
hope (related to climate and the future),
  2–4, 21, 55, 86, 116–17, 126, 133,
  152, 163, 223, 232

Hopkins, Michael S., 56
horror (film), 137–48
The Hungry Tide, 142, 181–82,
  193n2, 195n25

Ibis trilogy, 185–86, 192–93
Indigenous, 10–11, 14, 16–17, 122,
  184–85, 232
injustice(s): environmental, 4, 10; racial,
  14; in schools, 22
inquiry, 16, 31, 40, 96–97, 126–27, 161
insects, 31–36, 44, 107
Instagram, 4, 12–13, 15–20,
  23, 130, 156
Intergovernmental Panel on Climate
  Change (IPCC), 2, 67
International Space Station (ISS),
  50–51, 56–58, 64
"Iron Triangle," 190–91

Jah, Moriba, 59

Kane, Joan, 127
Kelly, Scott, 52–53, 62–64, 66
Kerridge, Richard, 79
Kessler Syndrome, 52
"Kiribati: The Face of Climate
  Change," 103
Kissock, Heather, 58
Krepon, Thomas, 55

Leeson, Craig, 125–26, 131
Leiserowitz, Anthony, 2
"Literature and Ecology: An Experiment
  in Ecocriticism," 78
The Lord of the Rings, 142
low Earth orbit (LEO), 51
"A Lullaby in Fracktown," 127

March for Our Lives, 15
Marsico, Katie, 64–65
Martinez, Xiuhtezcatl, 16–17, 123
Material Ecocriticism, 179; material
  turn, 179, 191, 192–93

Max Goes to the Space Station: A Science Adventure with Max the Dog, 56–57
May, Jamaal, 127
McKay, Adam, 1–3
"Mediating Climate Change: Ecocriticism, Science Studies, and *The Hungry Tide*," 142
medium Earth orbit (MEO), 51
Mellin, Lilace, 127
migration, climate, 177, 181, 183, 203, 213. *See also* climate refugee(s); displacement, climate-related
The Ministry of Utmost Happiness, 183, 195n34
misinformation, 38, 206, 209, 211
Moltz, James Clay, 54
My Journey to the Stars, 56, 62

Nakate, Vaness, 13
National Aeronautics and Space Administration (NASA), 2, 14, 22, 41, 44, 50, 52, 54–58, 60–61, 63–65, 95
National Oceanic and Atmospheric Association (NOAA), 60, 216–19
Native American, 14, 224. *See also* Indigenous
nature: dependence on, interconnectedness, 36, 55, 78–79, 143, 180; destruction, exploitation of, 1, 17, 76–77, 85, 89, 119, 140–41, 191; disconnection from, 33, 79–80, 232; representations of in film and literature, 77, 85, 138, 145–47, 183, 185, 189–90, 218
Nature's Children, 64
neoliberalism, 84
NOAA. *See* National Oceanic and Atmospheric Association
Nussbaum, Martha, 211
Nye, Naomi Shihab, 122

ocean acidification, 204–5

"Pacific Salt," 217
parking garage effect, 51, 53
participationism (sociocultural lens), 140
physics (class, lesson), 31, 38–39, 55, 231
plants, 65, 79, 106–7, 177, 189–90, 196; plant horror, 138, 144–48
A Plastic Ocean, 125, 126
Poems for a Small Planet: Contemporary American Nature Poetry, 127
poetry, 17–18, 122–23, 126–27
postcolonial ecocriticism. *See* ecocriticism
postcolonial green, 192
"Power Politics," 183, 186–87
Prestigiacomo, Max, 13–14

renewable energy, 97–99
Reset Earth, 81, 86–88
River of Smoke, 185, 189
robotics, 100, 102, 105–7
Rogoff, Barbara, 40
Ronald Reagan and the Public Lands: America's Conservation Debate, 1979–1984, 84
Roper, Kelly, 122
Roy, Arundhati, 177–93

Salgado, Stephanie, 13–14
Sargent, Mark, 41–42
satellites: communications and environmental monitoring, 49, 59, 61; health, navigation, 51, 58–61; safety and emergency management, 59–61; Terra, 50; weather forecasting, 58–61
School Strikes for Climate, 9
science (Earth, environmental, climate) 43, 49, 56–58, 65, 76, 79, 86, 96–97, 98–107, 122, 140, 143, 178, 204–13, 220–23, 231
science fiction, 55, 179, 216
scientific superiority complex, 43

scientific theory, 40, 42
The Sea of Poppies, 185
Sedwick, Raymond J, 53
Sentinel-6 Michael Freilich, 50
September 11 (9/11), 138, 141–48
Slovic, Paul, 80
social justice, 10, 96, 97, 109, 151
social media, 9–23, 90, 114,
    155–56, 201
space, 14, 22, 41, 51–67; space debris,
    49–67; space garbage nets, trucks,
    sails, 53, 61
space-age environmentalist, 57, 67
space literacy, 49–55
speculative fiction, 231–32
spoken word, 12, 16–18, 212
*Storylistening: Narrative Literacy and
    Public Reasoning*, 231
storytelling, 55, 212, 231–32; digital, 4,
    29–30; documentary, 29–31, 129
*The Storytelling Animal: How Stories
    Make Us Human*, 3
sustainability, environmental, 114–18;
    definition of, 152–54; in literature,
    193; teaching and frameworks, 32,
    38, 156–61

terrorism, 137, 140, 147
theory of mind (ToM), 206–7
Thunberg, Greta, 13–15, 20,
    123–24, 217
tidal barrage, 96–97, 98, 99
ToM. *See* theory of mind
traffic congestion, 96, 97, 100, 102

UN Environment Program, 86
Union of Concerned Scientists, 54
UN Office for Outer Space Affairs, 49

UN sustainable development goals
    (SDGs), 158
UN World Commission on Environment
    and Development, 152
urban planning, 58

video games, 3, 16, 76, 81, 86, 89–91
Vikalp Sangam project, 232
visual arts, 104

Wagner, Tony, 203
War on Terror, 138–47. *See
    also* terrorism
"Waterdevil," 127
*The Water Will Come: Rising Seas,
    Sinking Cities, and the Remaking of
    the Civilized World*, 219
Weik von Mossner, Alex, 2
Wells, H. G., 55
Wolfe, Tom, 63–64, 66
World Food Program, 36
"World of the Future, We
    Thirsted," 122–23
writing activities: film review, 31–32;
    informational, 104, 106; narrative
    prompts, 64–66; opinion and
    persuasive, 98, 125–26; skits,
    58–62; social media content, 2–13,
    18–20, 23–24
Writing the Climate, 79

Youth Climate Action Team
    (YCAT), 13, 15
youth climate movement, 4, 10,
    13–15, 19–20

Zaki, Jamil, 55–56
*Zobi and the Zoox: A Story of Coral
    Bleaching*, 204

About the Contributors

Beverly B. Bachelder is a long-time English language arts teacher, curriculum director, principal, adjunct professor, and currently workshop presenter with her husband, Robert Bachelder, on orbital space sustainability for such organizations as the Massachusetts Environmental Education Society, the Massachusetts Association of Science Teachers, the National Science Teachers Association, and STREAMS—an international conference highlighting exemplary practices in the environmental humanities. A graduate of Luther College, she earned her master's degrees from Yale University and the University of Kent at Canterbury, England. She was honored by the Massachusetts Blackstone Valley Education Foundation for her strong commitment to STEM education and career preparation, and in 2015, was named a recipient of the Priscilla B. Mason Arts and Culture Award for her leadership in science education. She worked collaboratively with the science teachers in her district to implement "Space Week" for grades six through eight—an interdisciplinary curriculum unit recognized for excellence by the New England League of Middle Schools.

Robert S. Bachelder is the retired president of a Massachusetts charitable organization that in 2009 launched a public education and advocacy program for a sustainable space environment. He subsequently authored a public resolution on the space debris issue that was adopted by a national civil sector institution and was a public commenter for a Federal Communications Commission proceeding on commercial satellite licensing rules for orbital debris mitigation. When he was in eighth grade and the modern environmental movement was gestating, he advocated with state legislators to clean up the Hudson River. Today he calls on environmental activists and green investors to expand the scope of their activities to include both Earth and space. A graduate of Dartmouth College (government) and Yale Divinity School (social ethics), his published essays and book chapters address subjects ranging from corporate ethics to nuclear deterrence.

Karen Ball, originally from Canada, has been working in the education field for the past seventeen years. She has worked and studied in a range of places, including Atlantic Canada, America, South Korea, Northern Canada, Austria, and currently in the Algarve, Portugal. She has held the roles of secondary school principal, primary school principal, and is currently deputy head of school. Her passions include special education, EAL, and online learning. She works to develop students through empowerment, physical activity, and growth mindset. Karen holds a master's degree in education and a leadership certification through Harvard University. She is currently working on her PhD at the University of Vienna in education science. She is an avid runner and has completed four full marathons to date.

Mary-Alice Corliss is a former high school science teacher and current science education and learning specialist at Cognia, who develops and reviews high-quality summative and formative science assessments for a wide range of stakeholders. She has past and current experience working directly with DOE leadership across the United States, as well as teachers, and district leaders, to best support students and teachers in their learning journey.

Elke de Vries, originally from Adelaide, Australia, has been working in the international education field for the past nine years. Spending most of her career in Vienna, Austria, she has held roles such as head of languages, careers coordinator, and humanities teacher. Elke has a love of the outdoors, the environment, and internationalism, something that she loves to bring to the classroom. She loves to drive change in education and make sure that students see that interconnectedness of the world around them. Instead of seeing issues or ideas as single, students are encouraged to develop a wholistic mindset when developing in the classroom and into their futures. Elke holds a master's degree in education and has currently moved "back home" to Adelaide and is looking forward to rediscovering the Australian nature with her family.

Carley Petersen Durden has been teaching composition since 2010. She is currently a member of the Language and Literature Department at Missouri Southern State University.

Jared Durden has been teaching college physics since 2013. He is currently a member of the Physical Science Department at Ozarks Technical Community College.

Erden El studied English language and literature at Ankara University, Turkey. He started teaching English in 2003. He took his MA degree from

Atılım University in English language and literature. He lived in Germany between 2015–2021, where he taught Turkish. He completed his PhD at Hacettepe University in the department of American culture and literature. He currently teaches English literature at Ankara Social Sciences University and Istanbul Aydın University.

Tatiana Konrad is a postdoctoral researcher in the Department of English and American Studies, University of Vienna, Austria, the principal investigator of "Air and Environmental Health in the (Post-)COVID-19 World," and the editor of the Environment, Health, and Well-being book series at Michigan State University Press. She holds a PhD in American studies from the University of Marburg, Germany. She was a visiting fellow at the University of Chicago (2022), a visiting researcher at the Forest History Society (2019), an Ebeling Fellow at the American Antiquarian Society (2018), and a visiting scholar at the University of South Alabama (2016). She is the author of Docu-Fictions of War: U.S. Interventionism in Film and Literature (2019), the editor of Cold War II: Hollywood's Renewed Obsession with Russia (2020) and Transportation and the Culture of Climate Change: Accelerating Ride to Global Crisis (2020), and a coeditor of Cultures of War in Graphic Novels: Violence, Trauma, and Memory (2018).

Alexandra Lakind is an artist, educator, and scholar working across an array of contexts conducting arts and educational programming to foster collaboration and environmental connection. Lakind has received formal training from Interlochen Arts Academy (HS), Royal Conservatoire of Scotland (BA), New York University (MA), and University of Wisconsin–Madison (PhD). Lakind's dissertation explores how children are positioned in the political project of addressing climate change and how social actions around climate change affect ideas about children. Lakind's related publications can be found in *Edge Effects* (the Center for Culture, History, and Environment's digital magazine) and *The International Journal of Press and Politics*. Lakind is a member of the collaborative project "In/Fertile Environments: Making Kin in a Time of Crisis" and is coeditor of the BECOMING series (becoming.ink).

Alexandra Laing is a well-respected educational leader working to promote and implement STEM and PBL education on a national level. In Alexandra's over fifteen years in education, she has been a teacher at the elementary and secondary levels, a school district K–12 STEM instructional specialist, a director of school leadership and accountability, and is currently partnered with 3DE by Junior Achievement as the national director of school leadership in which she oversees partnering 3DE public schools across the United States. Alexandra has presented at numerous national conferences on the implementation of standards-aligned problem-based learning, diversity, and equity through STEM in K–12 and higher education, collaborated with Magnet Schools of America and the Florida Department of Education for science and STEM education, and served as a 2019 Albert Einstein Distinguished Educator Congressional Fellow with Senator Jacky Rosen where she helped to develop and introduce multiple pieces of legislation and managed the Senator's education and STEM portfolios. Alexandra has a BS in elementary and middle education in English language arts and science, a MEd in educational leadership, and is currently pursuing her doctorate in educational leadership and administration at The George Washington University where she is a doctoral candidate working on her dissertation about the educational leader's perceived or actualized role in establishing grading practices that support student learning. She has been recognized and highly awarded for her contributions and dedication to education including being named the 2015 Boca Raton Rotary Club Sunrise Teacher of the Year, the 2015 J. C. Mitchell Elementary Teacher of the Year, the 2016 Florida Science Finalist for the Presidential Awards for Excellence in Mathematics and Science Teaching, and a 2020 ASCD Emerging Leader.

Helen Liu is a doctoral candidate in education at York University. Her research interests include the critical examination of media, adolescent development, and the study of international students. Her current research involves critically assessing discrepancies between the promotion and recruitment of international secondary student programs, and whether student experiences accurately reflect the standard that is advertised. Helen is also a teacher on the YRDSB, teacher assistant, and research assistant.

Alyssa Racco is a doctoral candidate in education at York University. She has a master's degree in education with an accompanying diploma in language and literacy. Her research interests include the education system's role in furthering speciesist ideals, sociolinguistics, and language acquisition. Her current research intends to explore the anthroparcal messages transmitted to students via mandated curriculum. Alyssa is also a teacher with the YCDSB, teacher assistant, research assistant, and project manager for a SSHRC Insight Grant.

Vandana Singh teaches physics at Framingham State University and works on a transdisciplinary, justice-centered pedagogy of climate change. She is also a writer of speculative fiction and a 2021 Climate Imagination Fellow at Arizona State University's Center for Science and the Imagination. Her writing website is http://vandana-writes.com/.

Suhasini Vincent defended her doctoral thesis on experimental writing within the postcolonial framework of Indian writing in English through a joint-supervision program between the University of Paris 3–Sorbonne Nouvelle and the University of Madras in 2006. In 2009, she was awarded the second prize for the best defended thesis on topics relating to the Commonwealth, organized under the aegis of the SEPC (Société d'études des Pays du Commonwealth). Her research focusses on the legal scope of environmental laws in postcolonial countries and explores ecocritical activism in the essays and literary works of Arundhati Roy, Amitav Ghosh, and Yann Martel. She is at present an associate professor at the University of Paris 2–Panthéon Assas since 2007.

Rebecca L. Young serves as a content manager for the nonprofit education organization Cognia and as an advisor for the International Baccalaureate's Middle Years Programme in Language and Literature Assessment. A former high school English teacher, her PhD work at SUNY Binghamton explored using literature as a lens for climate literacy. This is the focus of her first book, *Confronting Climate Crises through Education: Reading Our Way Forward*, and of her current research on how interdisciplinary instruction and assessment can prepare young people for the challenges of a changing world. She has contributed to *Critical Insights: Americans in Exile* and *Teaching Climate Change to Adolescents: Reading, Writing, and Making a Difference*. Her most recent publication, *Literature as a Lens for Climate Change: Using Narratives to Prepare the Next Generation*, offers a collection of chapters focused on the ways storytelling can engage K–12 students as ecologically conscious, globally minded problem solvers in the climate crisis.

www.ingramcontent.com/pod-product-compliance
Lightning Source LLC
Chambersburg PA
CBHW022307280326
41932CB00010B/1014